# Delay and Disruption Claims in Construction

## A practical approach

ice
Institution of Civil Engineers
publishing

# Delay and Disruption Claims in Construction

## A practical approach

Ali Haidar
Contracts Consultant, Dar Al Riyadh, Saudi Arabia

Peter Barnes
Contracts Consultant, Melbury Construction Consultants Ltd, Waltham Cross, UK

Published by ICE Publishing, 40 Marsh Wall, London E14 9TP.

Full details of ICE Publishing sales representatives and distributors can be found at: www.icevirtuallibrary.com/info/printbooksales

**Also available from ICE Publishing**

*Professional Service Agreements, Second edition.*
R. Barnes. ISBN 978-0-7277-4158-5
*ICE Manual of Construction Law.*
V. Ramsey, A. Minogue, J. Baster and M. O'Reilly (eds).
ISBN 978-0-7277-4087-8
*Client Best Practice Guide.*
Institution of Civil Engineers. ISBN 978-0-7277-3650-5
*Cost–Benefit Analysis, Second edition.*
M. Snell. ISBN 978-0-7277-4134-9

www.icevirtuallibrary.com

A catalogue record for this book is available from the British Library

ISBN: 978-0-7277-5705-0



ICE Publishing is a division of Thomas Telford Ltd, a wholly-owned subsidiary of the Institution of Civil Engineers (ICE).



FSC
www.fsc.org
MIX
Paper from responsible sources
FSC® C013604

Typeset by Academic + Technical, Bristol
Printed and bound in Great Britain by CPI Antony Rowe Limited, Chippenham and Eastbourne

# Contents

# Preface

Delay and disruption claims are common occurrences on construction projects. When they arise they need to be evaluated and managed effectively. There is often much debate about the approach that should be used when presenting or evaluating a delay and disruption claim. The purpose of this book is to serve as a legal and practical guide to the process of delay and disruption presentation and evaluation.

With this aim in mind, the chapters of this book deal with an understanding of construction and contract law (Chapter 2), breaches of contract and the remedies available (Chapter 3), delay and disruption issues (Chapter 4), and loss and expense (Chapter 5).

Chapters 6 to 8 contain case studies that deal with an extension of time and related losses claim, a loss and expense (and disruption) claim, and an increased costs claim, respectively.

This book not only sets out the legal basis of most claims, but also serves as a practical guide to the presentation of claims, based upon our collective years of general experience within the construction industry, and our specific experience of presenting and evaluating delay and disruption claims.

A key, and in some ways unique, element of this book is the recognition that in the international market, while the tried and tested methods of the UK construction industry are still frequently used, very often more dynamic approaches to the settlement of claims are used. Where claims can amount to many millions of US dollars, a painstaking and forensic approach to claims is often unnecessary and undesirable, and the parties may agree that a more pragmatic approach is more appropriate. This approach can prevent the parties from becoming involved in extensive and extended (and extremely costly) legal proceedings (perhaps stretching over several years), and enables disputes to be settled more rapidly and more easily. This allows the parties to continue with their normal business activities in a much shorter time than would have been the case had a more forensic exercise been undertaken. Therefore, very often in the emerging and rapidly growing international construction markets, global claims are accepted as a means of reaching a fair valuation of a claim, whereas in the more established UK market this approach is generally frowned upon other than in the most exceptional circumstances.

Although the construction industry is often considered to be a dispute-ridden industry, it is actually the case that the majority of construction projects are delivered within acceptable time and cost parameters. However, in a significant minority of construction projects this is not the case, and it is these projects that cause many of the disputes in the construction industry, and these disputes often centre around the responsibility for and the effect of delay and disruption events.

There is a great deal of, often conflicting, guidance on the subject of delay and disruption claims, and this is not helped by the courts, which appear to follow different approaches when considering delay and disruption matters. Consequently, it is important that practitioners understand the principles behind delay and disruption claims, and are aware of the various techniques that are available to carry out an assessment of those claims, so that an approach may be adopted that is appropriate to the circumstances of a particular case. Most delay and disruption claims are decided upon the facts of a particular case, rather than some overriding principle, and the important point is to ascertain the facts and then apply the law. The aim of this book is to provide the necessary legal analysis and information regarding the available delay and disruption analysis techniques to assist all of those involved in the construction process to be able to carry out the appropriate analysis and assessment of delay and disruption.

There is no one 'right' way to present a delay and disruption claim, as every claim is dependent on many interrelated issues that affect the way that a claim can or should be presented. There are, however, many pitfalls that may be faced in the preparation of a delay and disruption claim, and these pitfalls can often very easily be avoided if certain basic principles are understood and followed. It is hoped that this book will assist readers to follow the appropriate path in the preparation of their claims, and will help them to avoid the pitfalls along the way.

Finally, we are both most grateful to all those who, over the years, have contributed to our knowledge regarding the subject matter of this book, and we give special thanks to our families who have been a great support to us during the writing of this book.

# About the authors

**Dr Ali D. Haidar** has a PhD in Construction Management from South Bank University, a BSc and MSc in Civil Engineering from Oklahoma State University, and an MSc in Construction Law from King's College London.

Dr Haidar's experience in the construction industry spans 25 years, with much emphasis in recent years on contract management and claims. His knowledge extends to project management, artificial intelligence, modelling and, more recently, programme management. He is the author of many papers and has lectured at universities in the UK and the Middle East.

Dr Haidar has been involved in very large claims, totalling hundreds of millions of US dollars, and he has brought this experience to the writing of this book. His expertise in assessing and presenting claims has made him one of the major players and professionals in the Middle East.

Dr Haidar currently resides in Saudi Arabia with his family, and works as a contracts consultant at Dar al Riyadh. He is in the process of writing further books on global claims and contract management.

**Peter Barnes** is a Chartered Arbitrator and a UK Registered Adjudicator, and has an MSc in Construction Law and Arbitration from King's College London and a Diploma in International Commercial Arbitration from the Chartered Institute of Arbitrators.

He has been actively involved in the construction and civil engineering industry for over 35 years, with the last 10 years being spent dealing entirely with claims and dispute avoidance and/or resolution in that industry. Peter Barnes has acted as arbitrator, adjudicator, mediator and expert witness in both liability and quantum claims.

Peter Barnes is the author of the book *JCT'05 Standard Building Sub-Contract*; and is the co-author of *Subcontracting under the JCT 2005 Forms.*

He is based in the UK, near London, and is the co-owner of Melbury Construction Consultants Ltd, a practice specialising in claims in the construction and civil engineering industry.

**Delay and Disruption Claims in Construction**
ISBN: 978-0-7277-5705-0


doi: 10.1680/ddcc.57050.001

# Chapter 1
# Introduction

## 1.1. General
Construction is a unique industry, as it is a fast moving, complex and dynamic process that depends on the successful coordination of many separate and distinct parties to ensure the delivery of a project on time, within budget and to the required standard. This coordination is dependent on the application of proper planning, programming and project controls, used in conjunction with the applicable management techniques.

Risk is an inherent feature of construction and, as Sir Michael Latham noted in his report, 'no construction project is risk free. Risk can be managed, minimised, shared, transferred or accepted. It cannot be ignored' (Latham, 1994). As it is generally accepted that risk is inherent in construction, it must also be accepted that delay and disruption are inherent in the construction process and, as such, delay and disruption needs to be anticipated and managed in much the same way as risk is dealt with.

The occurrence of delay and disruption does not mean that the programming or management functions have failed, but it should be recognised as an event that needs to be properly managed so that its effects can be minimised where possible. This normally can only be achieved by a clear identification of the event that has or will cause the delay and/or disruption, and a proper analysis of the effect that the event will have on the progress of the works on site and/or the completion date for the project.

One of the most significant unanticipated costs in many construction projects is the financial impact associated with delay and/or disruption to the works. Assessing the impact of delay and disruption, and establishing a direct causal link from a delay or disruption event to the effect, liability and resulting damages can be an extremely difficult and complex exercise. In selecting the most appropriate technique for a particular project, the relevant facts, the timetable, the nature and number of delay and disruption events, as well as the size of the potential dispute must all be considered to ensure that the matter is dealt with proportionately.

## 1.2. The purpose of this book
In the modern and developing world it is becoming increasingly recognised that the construction process is simply part of a much larger process of procuring and obtaining a required facility (be that an airport, a hospital, a school, an office block or housing stock). The construction process is, therefore, often driven by commercial needs, and

the preparation and resolution of claims is consequently often dealt with on a more pragmatic basis than was sometimes the case in the past. While the reference to previous case law is essential to form an understanding of the legal principles behind most claims, it must be recognised that many of those court cases were decided at times when perhaps different economic and/or commercial pressures applied.

Today, in rapidly growing economies such as the Middle East and the Far East, claims are not always dealt with using the methods that were developed over many years in the UK. To deal with a claim of many millions of pounds (or US dollars, as is more often the case) by providing meticulous cost records, and cause and effect analyses of each and every event, would simply be too time consuming and costly for either party to contemplate. It is for this reason that, particularly in the rapidly growing economies, it is not uncommon for 'loss and expense' to be dealt with as contractual compensation based on preliminary and other allowances within the contract applied on a formula basis, and for 'global claims' to be applied. However, in the UK such an approach would only be considered in the most extreme of circumstances.

Thus the early chapters of this book deal in a more traditional way with the basic issues of construction and contract law, breaches of contract and remedies available, delay and disruption issues, and loss and expense, while the case studies in Chapters 6 to 8 deal with real-life situations and how claims are actually presented in the faster growing sectors of the world's economy. The approaches illustrated in these later chapters can prevent the parties to a dispute becoming involved in extensive and extended (and extremely costly) legal proceedings (perhaps stretching over several years), and enable disputes to be settled more easily. This allows the parties to resume their normal business activities much sooner than would be the case if a more forensic exercise had been undertaken.

The approach applied in the case studies presented combines the basic principles of law and contract that lie behind all well-prepared claims, and the more pragmatic approach that is now often used to reach a speedy, pragmatic and fair resolution of submitted claims in the modern world.

### REFERENCES

Latham M (1994) *Constructing the Team.* HMSO, London.

**Delay and Disruption Claims in Construction**
ISBN: 978-0-7277-5705-0


doi: 10.1680/ddcc.57050.003

# Chapter 2
# Construction and contract law

## 2.1. Construction law

In many ways, 'construction law' is no different to any other field of law. However, it is recognised as being a body of law that relates to those elements of the law that directly affect the construction and civil engineering industries. It covers things such as contract law, torts and construction claims, and it is particularly concerned with the effect that these elements of the law have on those participants (including employers, consultants, contractors and subcontractors) involved in the construction process.

Some parts of law apply only in respect of the relationship between particular people or bodies (this is referred to as civil law), while other parts apply to the rights or duties of a person or a body against the state (this is referred to as criminal law). This book deals only with civil law matters.

Furthermore, some law applies only in respect to the relationship between particular people or bodies (e.g. contract law), while other parts of law can have a much wider application (e.g. the tort of negligence). This book deals principally with contract law.

## 2.2. Contract law

The basic principles of contract law are outlined below.

A contract can be considered as being an agreement that gives rise to obligations which are enforced or recognised by law. Therefore the essence of any contract is *agreement*.

In deciding whether there has been an agreement, and what its terms are, a court usually looks for an *offer* to do, or to forbear from doing, something by one party, and an unconditional *acceptance* of that offer by the other party, turning the offer into a promise.

In addition, the law requires that a party suing on a promise must show that he has given *consideration* for the promise, unless the promise was given by *deed*.

Furthermore, it must be the intention of both parties to be legally bound by the agreement, the parties must have the capacity to make a contract, and any formalities required by law must be complied with.

If there is *fraud* or *misrepresentation*, the contract may be *voidable*, while if there is a *mutual mistake* about some serious fundamental matter of fact this may have the effect of making the contract *void*.

Finally, there must be sufficient *certainty of terms*.

Each of these elements of the contract are considered further below.

## 2.3. 'Offer and acceptance'

As stated above, for there to be an agreement, the very starting point is an offer being made by one party and an acceptance of that offer by the other party. This is not always as simple as it sounds, particularly when there have been prolonged negotiations between the parties. The courts have developed their own rules for looking at such matters.

In terms of offer and acceptance, the courts adopt what is sometimes known as the 'mirror image' rule of contract formation. That is to say that there must be a clear and unequivocal offer, which is matched by an equally clear and unequivocal acceptance of that offer.

An offer is a statement by one party of a willingness to contract on definite stated terms and is intended to be legally binding provided that those terms are, in turn, unequivocally accepted by the party or parties to whom the offer is addressed.

There is generally no requirement that the offer be made in any particular form; it may be made orally, in writing or by conduct. Of course, if a dispute were to arise in the future, then it would be beneficial for the offer to be in writing.

In whichever form an offer is made, it must be sufficiently definite to be capable of resulting in a contract if accepted. Its terms and conditions must be clear and unequivocal, and it must be made with the intention that it is to become legally binding as soon as it is accepted by the person to whom it is addressed. In this context, a person includes a corporation because, in law, a corporation is a legal person; that is to say, a corporation is regarded by the law as a legal entity quite distinct from the person or persons who may for the time being be the member or members of the corporation.

For agreement to be reached there must be a clear and unequivocal acceptance to a clear and unequivocal offer. The acceptance must be unqualified (i.e. as noted earlier it must 'mirror' the offer). As a general rule, silence does not constitute acceptance (*Felthouse* v. *Bindley* 1862), and neither does inactivity. Given this, the general rule is that an acceptance has no effect until it is communicated (either in writing or orally) to the party making the offer. The main reason for this is that it could cause hardship to the party making the offer if he were bound without knowing that his offer had been accepted.

If the acceptance does not clearly and unequivocally accept the offer (e.g. it seeks to add to or vary the terms contained in the offer), then it is simply a counter-offer (not an acceptance), and a counter-offer destroys the original offer. A counter-offer has the same status as an offer in the formation of a contract, and consequently a counter-offer must be clearly and unequivocally accepted before agreement has been reached.

In respect of construction works, in particular, there is frequently a whole series of counter-offers made before an acceptance is given. This may be because the parties are negotiating about the terms, or because they are each trying to impose their own terms. This latter situation is often called the 'battle of the forms', and variants of this battle of the forms are at the base of many disputes between contractors and subcontractors.

## 2.4. Agreement

As noted previously, the essence of any contract is agreement. The test for the existence of an agreement is objective rather than subjective. In other words, the existence is tested on the facts, rather than on what may have been perceived to be the intention of the parties. The principal justification for the adoption of this test is the need to promote certainty.

An agreement is reached either

- when a statement of agreement is signed, or
- when one party makes an unambiguous offer that can be accepted, and the other party accepts it unequivocally.

## 2.5. Consideration

Other than where a contract is executed as a deed (see below), an agreement requires consideration to be exchanged between the contracting parties before it becomes binding. The classic definition of consideration was expressed in *Currie* v. *Misa* (1875) LR 10 Ex 153, in the following terms:

> a valuable consideration, in the sense of the law, may consist either in some right, interest, profit or benefit accruing to the one party, or some forbearance, detriment, loss or responsibility given, suffered or undertaken by the other.

In the ordinary building contract situation, the consideration given by the employer is the price paid or the promise to pay, and the consideration given by the contractor is the carrying out of the works or the promise to carry them out.

## 2.6. Executed as a deed

As noted above, consideration is not required in the case where a contract is executed as a deed. Contracts, entered into under hand (meaning by individual signature of a duly authorised person) are termed 'simple contracts'. With such contracts the limitation period for claims is 6 years from the breach. However, if a contract is entered under deed (meaning by corporate seal countersigned by authorised persons, or by signature of the company secretary and another director), the limitation period for claims is extended to 12 years from the breach.

## 2.7. Fraud or misrepresentation

A misrepresentation may be defined as an unambiguous, false statement of fact, which is addressed to a party and which misleads that party, and which materially induces that party to enter into a contract.

Whether made fraudulently or innocently, the general effect of such a misrepresentation is to render the contract voidable. As a general rule, simple silence is not an example of misrepresentation (*Keates* v. *Cadogan* (1851) 10 CB 591).

The principal remedies for misrepresentation are rescission (i.e. revoking the contract) and damages.

## 2.8. Mutual mistake

This is the position where the parties to the contract are mutually mistaken about some serious material fact fundamental to the contract which may render the contract void.

## 2.9. Void or voidable

A contract that is *void* is a contract that has no legal effect (i.e. it is a nullity), as distinct from a contract that is *voidable*, which is simply a contract that can be set aside (or rescinded) by one of the parties.

## 2.10. Certainty of terms

Even if there is clear agreement through offer and acceptance, a contract may fail to come into existence because of uncertainty as to what has been agreed (*Scammell* v. *Ouston* (1941) 1 All ER 14).

Although it has been found that 'the parties are to be regarded as masters of their contractual fate in determining what terms are essential' (*Pagnan* v. *Feed Products* [1987] 2 Lloyds Rep 601, CA), and 'It is for the parties to decide whether they wish to be bound and, if so, by what terms, whether important or unimportant' (*Mitsui Babcock* v. *John Brown Engineering* (1996) CILL 1189), it is generally considered that agreement as to the parties, price, time and description of works is normally the minimum necessary to make the contract commercially workable.

## 2.11. Standard forms of contract

Contracts can be formed in many different formats, and may be produced on an *ad hoc* basis for each and every agreement between the parties. Alternatively, a contract may be formed using a standard form of contract.

Standard forms of contract can be traced back to the nineteenth century (if not earlier). However, it appears to be only fairly recently, largely as a result of the Latham Report (1994), that standard contract forms have come to be seen as having a dual purpose

- to set out the rights and obligations of the parties, and
- to ensure that the risk of the project is allocated to the party that can best manage that risk.

In addition to these functions, there are some obvious benefits in using standard forms of contract. These include

- there is no need to produce (and incur the legal costs of producing) *ad hoc* contracts for every project
- there is a degree of certainty regarding the interpretation of the clauses of the contract (particularly for those standard forms that have been in existence for some time, and for which some of the more important clauses have been tested in the courts)
- the parties know (with reasonable certainty) the consequences of various possible courses of action.

There are many standard forms of contract in use in the construction and civil engineering industries, and a brief commentary on some of the more commonly used standard forms is provided below.

### 2.11.1 NEC 3

NEC 3 is a generic name for a family of contracts published for the Institution of Civil Engineers by Thomas Telford Limited. NEC stands for New Engineering Contract, and it is by this acronym that the contracts are generally known. The main contract and the subcontract were first published as consultative editions in 1991. The first formal editions were issued in 1993, second editions in 1995, and third editions in 2005.

It was always intended that there would be a family of New Engineering Contracts, and between 1991 and 2005 other contracts were produced. By 2005 the NEC 3 family of contracts comprised

- the NEC 3 Engineering and Construction Contract
- the NEC 3 Engineering and Construction Subcontract
- the NEC 3 Professional Services Contract
- the NEC 3 Short Contract
- the NEC 3 Short Subcontract
- the NEC 3 Adjudicator's Contract
- the NEC 3 Term Services Contract
- the NEC 3 Framework Contract.

The contracts are supported by officially published guidance notes, flow charts and the advisory document *NEC 3 Procurement and Contract Strategies*.

The New Engineering Contract was drafted with the objectives of achieving flexibility, stimulus to good project management, clarity and simplicity.

- *Flexibility*. The NEC 3 Engineering and Construction Contract aims to be an all-purpose contract for all construction and engineering disciplines in the UK and internationally. It offers this through a combination of uniquely drafted provisions and a complex structure of options. The main options comprise six types of payment mechanism: priced contract with activity schedule (option A); priced contract with bill of quantities (option B); target contract with activity schedule (option C); target contract with bill of quantities (option D); cost reimbursable contract (option E); and management contract (option F). The secondary option clauses are labelled with prefixes X and Z. Not all such clauses would normally be considered as being secondary. Included within them are some matters, such as retention and liquidated damages for late completion, which most other contracts treat as essential. Other matters, such as performance bonds and performance-related damages (together with amendments to the contract by way of the unspecified Z clauses), are more obviously contract specific.

- *Stimulus to good management.* A principle of the NEC 3 is the need to concentrate as much on management as on the obligations and liabilities of the parties. Therefore, NEC 3 contracts lay great emphasis on communications, cooperation, programming, early warnings, risk registers, and the need for clear definition at the outset of various types of information.
- *Clarity and simplicity.* The NEC 3 is written in non-legalistic language using short sentences, and avoids cross-references.

### 2.11.2 ICE Conditions of Contract

This is a family of standard conditions of contract for civil engineering works. The Institution of Civil Engineers (ICE) is an independent engineering institution representing approximately 80 000 civil engineers worldwide. Its principal membership is in the UK, but it has members in China, Hong Kong, Russia, India and roughly 140 other countries. The *ICE Conditions of Contract*, which have been in use for over 50 years, were designed to standardise the duties of contractors, employers and engineers, and to distribute the risks inherent in civil engineering to those best able to manage them.

The first edition of the *ICE Conditions of Contract* was launched in December 1945, and the subsequent editions have become one of the main forms of standard contract for UK civil engineering work.

Pre-1987 versions of Fedération Internationale des Ingénieurs-Conseils (FIDIC) contracts (see below) were very much based on the ICE forms, but there is far less similarity now. Traditionally, the ICE contracts are engineer-based contracts, where the employer has appointed the engineer.

The contracts presently in use include

- *ICE Conditions of Contract, Measurement Version* (7th edition, July 2004)
- *ICE Conditions of Contract, Design and Construct* (2nd edition, July 2004)
- *ICE Conditions of Contract, Minor Works* (3rd edition, July 2004)
- *ICE Conditions of Contract, Term Version* (July 2004)
- *ICE Conditions of Contract, Ground Investigation* (2nd edition, July 2004).

The ICE 7th edition is administered by an engineer who is appointed by the employer and acts as his agent, but is required to act fairly towards the contractor.

### 2.11.3 Joint Contracts Tribunal (JCT) contracts

The Joint Contracts Tribunal (JCT) was established in 1931 and for 75 years has produced standard forms of contracts, guidance notes and other standard documentation for use in the construction industry, principally in the UK. In 1998, the JCT became incorporated as a company limited by guarantee, and commenced operation as such in May 1998.

Currently, JCT forms require the agreement of eight constituent bodies before they are issued by the JCT. The constituent bodies are intended to be reasonably representative of the interests across the UK construction industry (namely, the employers, the consultants, the contractors and the subcontractors), and the JCT forms are naturally a reflection of these competing interests.

A benefit of the JCT Forms of Contract is that they are tried and tested, and there is, therefore, a great deal of certainty regarding the meaning of various clauses and how the courts will interpret those clauses (particularly as many of the clauses have at some time or another been considered by the courts). The perceived problem with the JCT Forms of Contract is that they are cumbersome and unwieldy, and are not flexible enough to cope with modern requirements. However, many of these perceived problems were dealt with by the publication in 2005 of an entirely new suite of contracts and subcontracts.

Some of the more commonly used JCT contracts are the

- Standard Building Contract
- Intermediate Building Contract
- Minor Works Building Contract
- Design and Build Contract
- Constructing Excellence Contract.

### 2.11.4 Fédération Internationale des Ingénieurs-Conseils (FIDIC) contracts

The Fédération Internationale des Ingénieurs-Conseils (FIDIC) produces standard forms of contract for civil engineering construction which are used throughout the world. FIDIC contracts are often referred to as the international standard.

FIDIC is the leading body for the development of model standard forms of contract for use in the international construction industry. The standard forms are generally accepted by employers and contractors as providing a balanced allocation of risks and providing fair procedures for administration of contracts.

The present suite of FIDIC contracts, often best known by their colours, replaces the original set issued in 1987. The latest contracts are

- Conditions of Contract for Construction (1st edition, 1999), The Red Book
- Conditions of Contract for Plant and Design–Build (1st edition, 1999), The Yellow Book

- Conditions of Contract for EPC/Turnkey Projects (1st edition, 1999), The Silver Book
- Short Form of Contract (1st edition, 1999), The Green Book.

Other Forms of FIDIC contract are

- Form of Contract for Dredging and Reclamation Works ('Dredgers' Contract'), The Blue Book.

In addition, FIDIC publishes

- a form of agreement for the engagement of consultants (The White Book).

The new FIDIC forms have adopted much of the current thinking around the administration of contracts, and have attempted to balance the risk and responsibility between the employer and the contractor.

## Cases

*Currie* v. *Misa* (1875) LR 10 Ex 153
*Felthouse* v. *Bindley* (1862)
*Keates* v. *Cadogan* (1851) 10 CB 591
*Mitsui Babcock* v. *John Brown Engineering* (1996) CILL 1189
*Pagnan* v. *Feed Products* [1987] 2 Lloyds Rep 601, CA
*Scammell* v. *Ouston* (1941) 1 All ER 14

REFERENCES

Latham M (1994) *Constructing the Team.* HMSO, London.
*NEC 3 Procurement and Contract Strategies* (2009) Thomas Telford, London.

**Delay and Disruption Claims in Construction**
ISBN: 978-0-7277-5705-0


doi: 10.1680/ddcc.57050.011

# Chapter 3
# Breach of contract and remedies

## 3.1. Breach of contract

A breach of contract is committed when a party without lawful excuse fails or refuses to perform what is required of him under a contract. A breach of contract does not automatically bring a contract to an end. However, a breach of contract gives to the innocent party a right to claim damages, and it may give him the additional right to terminate performance of the contract. The right to terminate performance of a contract may be by way of an express provision in the contract, or may be because the breach of the contract is of such a fundamental and repudiatory nature that it is considered to be a repudiatory breach of the contract or (if it is stated as being an intention in the future) an anticipatory repudiatory breach of the contract. This chapter considers in more detail damages arising from a breach of contract.

## 3.2. Damages

The purpose of damages is to compensate the innocent party for the loss which he has suffered as a result of the breach of contract by the other party. The aim of an award of damages is to put the innocent party in the same position, insofar as is possible in monetary terms, that he would have been in had the other party performed its obligations under the contract.

The innocent party must take reasonable steps to mitigate his loss. When pursuing a claim for damages, it must be shown that there was a term of the contract that a party breached (i.e. failed to comply with), and that the damages incurred arose directly as a result of that breach of contract.

Damages cannot be recovered where the loss that the innocent party has suffered is too remote a consequence of the other party's breach of contract. The general principle is that the innocent party can only recover losses that were within the reasonable contemplation of the parties at the time when the contract was entered into.

The famous case of *Hadley* v. *Baxendale* (1854) 9 Exch 341, is often cited to show that losses which occur 'naturally' or as a 'usual course of things' can be recovered as damages (this is usually referred to as the 'first limb' of *Hadley* v. *Baxendale*), and that losses that do not occur 'naturally' but which were in the reasonable contemplation of the parties at the time when the contract was made can also be recovered (normally referred to as the 'second limb' of *Hadley* v. *Baxendale*).

The details of the *Hadley* v. *Baxendale* case demonstrate the above points. Hadley was the proprietor of a mill in Gloucester, England. A crankshaft in the mill broke and Baxendale was employed to take the crankshaft to Greenwich, England, by a certain date for repairs. However, Baxendale failed to deliver the crankshaft at the due date, and this caused Hadley to lose business. Hadley sued for the profits he lost due to his loss of business. At first instance, the court found that Hadley could recover his loss of profits, and Baxendale was ordered to pay damages of £25. Baxendale appealed against this decision, contending that he did not know that Hadley would suffer any particular damage by reason of the late delivery, and at the appeal the court declined to allow Hadley to recover lost profits in this case, holding that Baxendale could only be held liable for losses that were generally foreseeable, which may have been the case if Hadley had mentioned his special circumstances in advance. The court decided that the mere fact that a party is sending something to be repaired does not indicate that they would lose profits if it were not delivered on time.

Although the two limbs of *Hadley* v. *Baxendale* are often cited, more accurately the test is simply a test of remoteness of damages. Under this test, if the damages claimed are too remote from the breach of contract then, unless that particular circumstance was in the contemplation of the parties at the time when the contract was formed, then those particular damages cannot be recovered.

### 3.2.1 Date of assessment of damages

The general rule is that damages should be assessed as at the date when the cause of action arises (*Miliangos* v. *George Frank* [1976] AC 443, HL; *Dodd Properties* v. *Canterbury City Council* [1980] 1 WLR 433, CA), but there are many exceptions, and it has been said that 'this so-called general rule . . . has been so far eroded in recent times . . . that little of practical reality remains of it' (Ormrod L.J. in *Cory & Son* v. *Wingate* (1980) 17 BLR 104, CA). Thus 'where it is necessary in order adequately to compensate the plaintiff for the damage suffered by reason of the defendant's wrong a different date of assessment can be selected' (Lord Browne-Wilkinson in *Smith New Court* v. *Scrimgeour Vickers* [1997] AC 254, HL).

### 3.2.2 Measure of damages

As noted above, the purposes of damages is to put the innocent party (i.e. normally the claimant), 'so far as money can do it' (*Robinson* v. *Harman* (1848) 1 Ex 850), back to the same position they would have been in if the contract had been properly performed.

However, sometimes the proper measure of damages is not the cost of reinstatement but the difference in value between the work actually produced and the work that should have been produced (*Dodd Properties* v. *Canterbury City Council* [1980] 1 WLR 433, CA). This will particularly be the case where the claimant has no prospect or intention of rebuilding, or where it would be unreasonable to award the cost of reinstatement.

The frequently quoted UK House of Lords case of *Ruxley* v. *Forsyth* [1995] 3 WLR 118 involved a swimming pool that was not constructed to the required depth. The swimming pool should have been constructed to a depth of seven feet and six inches, but when the pool was completed its depth was only six feet and nine inches. Forsyth sought damages for the reinstatement costs (i.e. to rebuild the swimming pool so that it was seven feet and six inches deep). However, the House of Lords found that it would be unreasonable for the claimant to insist on reinstatement because the cost of the work involved would be out of all proportion to the benefit obtained, and the claimant's measure of damages would simply be the difference in value (sometimes referred to as the 'loss of amenity value').

## 3.3. Liquidated damages

An alternative to pursuing a claim for damages is to insert into the contract a clause which states the amount of damages that will be payable in the event of a breach of contract. Such a clause is referred to as a 'liquidated damages clause'. In construction contracts, liquidated damages are most often used in respect of a contractor not completing the works on time and, because of that breach, the employer seeks to recover liquidated damages on the basis of a certain fixed financial amount per week (or part thereof).

A liquidated damages clause helps to eliminate uncertainty because the parties know in advance the financial liability that will arise if the contractor fails to complete the works on time. This acts as a protection for the contractor against unliquidated general damages claims, and enables the contractor to fix the price for risk. For an employer, a liquidated damages clause avoids the difficulty, time and expense involved in proving and assessing the actual loss that a party will suffer in the event of a breach.

The courts have found that a liquidated damages clause is only enforceable if it represents a genuine pre-estimate of the losses that are likely to be incurred as a result of breach of contract (*Regional Construction Sdn. Bhd.* v. *Chung Syn Kheng Electrical Co. Bhd.* [1987] 2 MLJ 763).

There are a number of grounds on which a liquidated damages clause may be held to be unenforceable, including ambiguity, its penal nature, the failure to provide for an extension of time in cases where the employer prevents or delays completion, and the failure to comply with contractual procedures (*Bramall & Ogden Limited* v. *Sheffield City Council* (1985) 1 Con LR 30). However, the effect of attacking the validity of a liquidated damages clause is, depending on the wording of the contract, and perhaps the reason for the unenforceability of the clause, to remove the ready-made and pre-set damages mechanism and substitute it with a liability to pay general or unliquidated damages, which could, subject to proof and ascertainment, be significantly higher than the liquidated damages rate.

### 3.3.1 Ambiguity

As a general rule, liquidated damages clauses must be constructed strictly *contra proferentem*. Therefore, where the clause is ambiguous and all other methods of construction have failed to resolve the ambiguity, the court may construe the words against the party seeking to rely on the clause.

### 3.3.2 Penal nature

A liquidated damages clause that is found to be penal is generally invalid. It is a well-accepted principle that liquidated damages provisions must, in both their amount and operation, constitute a genuine pre-estimate of the loss that the employer is likely to suffer as a result of the breach. It should be noted that the pre-estimate does not need to bear any relationship to the actual damage suffered as a consequence of the breach. If a liquidated damages clause is found to be penal in nature, the clause will be invalid and unenforceable, and the court will only enforce the sum identified where it represents a proper pre-assessment of the loss. The distinction between liquidated damages and a penalty is a matter of construction for the court, and it is for the person from whom the damages are claimed to show that the clause is, in fact, a penalty.

The use of the words 'penalty' or 'liquidated damages' in the contract is not conclusive or determinative of whether the liquidated damages clause is, in fact, one or the other.

The provisions to consider when addressing penalties with regard to liquidated damages in construction contracts can be summarised by the following two criteria.

- The essence of a penalty is a payment of money stipulated as against the offending party; the essence of liquidated damages is a genuine covenanted estimate of damages to the innocent party. The key is to attempt to estimate what actual loss would be caused by the breach by the offending party and, furthermore, whether it is fair and reasonable.
- The question of whether a sum stipulated is a penalty or is liquidated damages is a question of construction to be decided upon the terms and inherent circumstances of each particular contract, judged at the time of the making of the contract, not at the time of the breach.

To assist in this task of construction of such provisions, various tests have been suggested in respect of purported liquidated damages.

- It will be held to be penalty if the sum stipulated for is extravagant and unconscionable in amount in comparison with the greatest loss that could conceivably be proved to have followed from the breach.
- It will be held to be a penalty if the breach consists only in paying a sum of money, and the sum stipulated is a sum greater than the sum which ought to have been paid (i.e. greater than the actual losses of the innocent party).
- There is a presumption that it is penalty when a single lump sum is made payable by way of compensation, on the occurrence of one or more or all of several events, some of which may occasion serious and others but trifling damage.
- The time for assessment or construction of the provision as either a genuine pre-estimate of damages or a penalty is as at the time of the contract. In other words, if it is a genuine pre-estimate of damages or losses of the owner at the time the contract is entered, then it will likely be valid and enforceable.

In the case of *Commissioner of Public Works* v. *Hills* [1906] AC 368, Lord Dunedin formulated the test for penalty clauses as follows:

> The general principle to be deduced is that the criterion of whether a sum, be it called penalty or damages, is truly liquidated damages, and as such not to be interfered with by the court, or is truly a penalty which covers the damage if proved, but does not assess it, is to be found in whether the sum stipulated for can or cannot be regarded as a 'genuine pre-estimate' of the creditor's probable or possible interest in the due performance of the principal obligation.

The various tests of causation to assist in identifying a penalty as opposed to damages have been established again by Lord Dunedin in *Dunlop Pneumatic Tyre Co.* v. *New Garage & Motor Co. Ltd* [1915] Ac 79. Lord Dunedin reached the conclusion that, if the sum to be paid for a breach of the contract was substantial and arbitrary and bore no relation to the potential loss of the other party, it was a penalty. He stated:

> The essence of a penalty is a payment of money stipulated as *in terrorem* of the offending party; the essence of liquidated damages is a genuine covenanted pre-estimate of damage. The question whether a sum stipulated is penalty or liquidated damages is a question of construction to be decided upon the terms and inherent circumstances of each particular contract, judged of as at the time of the making of the contract.

In *Robophone Facilities Ltd* v. *Blank* [1966] 1 WLR 14, this prohibition of the penalty clause is said to be consistent with the law in respect of the treatment of damages, as damages are compensatory, and that to allow a clause which allows recovery of damages in excess of the actual loss suffered or sufferable would be wrong. The damages would be more of a deterrent, designed to discourage breaches of contract or to secure performance by the contractor, than compensation. Lord Justice Diplock stated:

> The onus of showing that such a stipulation is a 'penalty clause' lies upon the party who is sued upon it. The terms of the clause may themselves be sufficient to give rise to the inference that it is not a genuine estimate of damage likely to be suffered but is a penalty. Thus it may seem at first sight that the stipulated sum is extravagantly greater than any loss which is liable to result from the breach in the ordinary course of things. This would give rise to the *prima facie* inference that the stipulated sum was a penalty. But the plaintiff may be able to show that owing to special circumstances outside 'the ordinary course of things' a breach in those special circumstances would be liable to cause him a greater loss of which the stipulated sum does represent a genuine estimate.

In *Philips Hong Kong* v. *A-G of Hong Kong* [1993] 61 BLR 41, Lord Woolf was of the opinion that the courts should not try too hard to find a penalty clause; for example, the identification of circumstances where the liquidated damages would amount to an over-recovery by the plaintiff. The key question remains whether the liquidated damages were a genuine pre-estimate of loss at the time they were set. He stated the following:

> Except possibly in the case of situations where one of the parties to the contract is able to dominate the other as to the choice of the terms of a contract, it will normally be insufficient to establish that a provision is objectionably penal to identify situations where the application of the provision could result in a larger sum being recovered by the injured party than his actual loss. Even in such situations so long as the sum payable in the event of non-compliance with the contract is not extravagant, having regard to the range of losses that it could reasonably be anticipated it would have to cover at the time the contract was made, it can still be a genuine pre-estimate of the loss that would be suffered and so a perfectly valid liquidated damage provision. The use in argument of unlikely illustrations should therefore not assist a party to defeat a provision as to liquidated damages.

### 3.3.3 Failure to provide for an extension of time in cases where the employer prevents or delays completion

In a case where a contract does not contain a proper mechanism for allowing the employer to grant an extension of time for delays that are caused by the employer or by anyone or anything which the employer is responsible for, liquidated damages are irrecoverable, as the revised completion date cannot be set or identified under the provisions of the contract. In other words, time becomes 'at large', through the operation of the prevention principle (the concept of 'time at large' and the prevention principle are dealt with in more detail in Chapter 4). In such a case, the employer cannot impose a liquidated damages clause. The leading authority on this point is the decision in *Peak Construction (Liverpool) Ltd* v. *McKinney Foundations Ltd* (1970) 1 BLR 111. In that case, Lord Justice Salmon confirmed that a liquidated damages clause could not apply, and further commented:

> I cannot see how, in the ordinary course, the employer can insist upon compliance with a condition if it is partly its own fault that it cannot be fulfilled.

In the Australian case of *Gaymark Investments Pty Limited* v. *Walter Construction Group Limited* [1999] NTSC 143, the contractor (Walter) was delayed in completing the work by causes for which the employer (Gaymark) was responsible and which constituted acts of prevention by Gaymark. Under the provisions of the contract, Gaymark was not able to grant an extension of time to Walter, nor was it able to set a revised completion date, and, as a result of this, time became at large and Walter was simply obliged to complete the work within a reasonable time. In his judgment, J. Bailey, affirmed that, as a result of the circumstances, Gaymark was not entitled to deduct liquidated damages and time was set at large based on Gaymark's act of prevention. J. Bailey said:

> In the circumstances of the present case, I consider that this principle [i.e. the 'prevention principle'] presents a formidable barrier to Gaymark's claim for liquidated damages based on delays of its own making... I do not consider that there was any manifest error of law on the face of the award or any strong evidence of any error of law in the arbitrator holding that the prevention principle barred Gaymark's claim to liquidated damages.

### 3.3.4 Failure to comply with contractual procedures

Obviously, the contract should both reflect the employer's expectations and be consistent throughout, otherwise there could be serious consequences with regard to the validity of the liquidated damages provisions. A typical example of this is the employer's desire to take possession or part occupation as soon as possible, which, unless the contract specifically allows for this to happen, means that the employer could be risking his right to the recovery of liquidated damages should the contractor subsequently be late. This is best illustrated by the case of *Bramall and Ogden Limited* v. *Sheffield City Council* (1985) 1 Con LR 30. The facts of this case were that the Council had employed Bramall under a JCT 1963 standard form of contract to construct 123 houses. Within the appendix, the Council had inserted the following for liquidated and ascertained damages for late completion: 'at the rate of £20 per week for each uncompleted dwelling'.

During the course of the project, the Council decided to take possession of the houses as they were completed, with the architect issuing practical completion certificates.

However, the contract did not allow for this, as the parties had not agreed on or entered into a sectional completion supplement. The architect also provided various extensions of time, giving an extended completion date of 4 May 1977, with the last few houses being accepted on 29 November 1977. With the project being under a delay of around 7 months, the Council understandably decided to apply liquidated damages. This placed the Council in a quandary. Did they deduct for a proportion of houses which were not completed, or did they go for all 123 dwellings, as after all this were the works late? The Council opted for the latter option, and made a deduction for all 123 dwellings.

On a court appeal from an arbitrator's decision, Bramall successfully argued *inter alia* that there was a discrepancy in calculating liquidated damages because, as they were expressed per dwelling, this could not be reconciled with partial possession as this related to the works as a whole (i.e. not to individual dwellings); and, as liquidated damages must be applied, in accordance with the appendix of the contract, to the whole of the Works, the liquidated damages provision must fail. This would still be applicable, even if the Council had opted for not deducting all the damages but had decided on deducting damages on a *pro rata* basis of the works completed.

## 3.4. Loss and expense

Because, through the operation of construction contracts, breaches of contract often arise (e.g. the late completion of the works, for a variety of possible reasons), the right to recover loss and expense is included in many construction contracts as being the means of recovering losses arising from a specified breach (or breaches) of contract. In some ways, loss and expense can be considered as being a contractual mechanism for the recovery of what may otherwise be considered as damages.

Most often, loss and expense claims are pursued by a contractor for losses incurred as a result of the prolongation and/or disruption of the works.

Loss and expense claims have certain advantages over common law damages claims. Some of these advantages are

- there is a clear definition of responsibilities and procedures
- there is a clarification of the events and the circumstances for claims to be made
- there is a right to interim payments and/or decisions
- there is the avoidance of the uncertainties involved in claiming at common law
- loss and expense claims create a right under the contract to a debt, rather than a claim for damages.

Normally, the inability to recover monies under the express terms of the contract for reasons such as not conforming with the procedures required, does not preclude a claim at common law on the same matters, provided there is the entitlement at common law to claim damages, and provided that the contractor's common law rights have not been specifically excluded by the terms and conditions of the contract.

If a matter stated in the contract as giving rise to an entitlement to loss and expense is a breach of contract, then the contractor will normally have a right to common law damages. An example of this would be the late issue of information by the contractor.

However, if the matter relied on does not amount to a breach of contract (e.g. issuing variations), then common law damages may not be applicable.

On the other hand, the preservation of the contractor's common law rights may entitle the contractor to make common law damages claims in respect of matters for which there are no express provisions within the loss and expense clauses.

The authority on the application of damages arising from the contract, both in addition to and, where appropriate, instead of, the express provisions of the contract for loss and expense, is clearly stated in the case of *Stanley Hugh Leach* v. *Merton* (1985) 32 BLR 51.

It is important to note that, should the contractor need to pursue his claim at common law because he has not followed the procedures under the contract, then it is possible that he may be penalised by the court if his non-conformity with the contract procedure did not allow the employer to mitigate his loss.

In some circumstances (albeit relatively uncommon), the parties may seek to exclude (via a contract term) the parties' common law rights, so that only the terms of the contract itself govern the relationship between the parties. However, it must be remembered that the courts view this course of action with some caution, and consequently there must be a very clearly worded term in the contract excluding the parties' common law rights for it to be effective.

Most claims, whether for time or money, involve establishing what is often called the 'nexus of cause and effect' (i.e. the link between cause and effect). This means that what needs to be proved is that, because of the occurrence of a particular event, certain things happened, and as a direct result of those events happening this in turn led to one of the parties incurring delays or costs which were not previously contemplated by the parties, and which it would not have been reasonable for the parties to have contemplated.

In order to establish this nexus (or link), which is never an easy task in a construction situation, records need to be available that show that the circumstances which existed before an event occurred changed after that event occurred, and that that change in circumstances could only have been as a result of the event in question.

A claim for loss and/or expense is the means of putting a contractor back into the position in which he would have been but for the delay or disruption; it is not a means of turning a loss into a profit (unless, of course, the loss is because of the non-payment of loss and/or expense). Therefore, where a contractor's costs are materially different from the rates quoted in the contract for preliminaries, there must be a good reason for that difference, otherwise the claimed costs may be disallowed in total or in part on the grounds that the costs claimed are unreasonable.

The settlement in effect mirrors a common law damages claim, and therefore an exact establishment of the contractor's additional costs must be made. At first sight it might be thought that the establishment of those additional costs is relatively easy, but in fact it can be very difficult. There is an increasing trend in the growing economies of the Middle East and the Far East (in particular) for a more formulaic approach to this exercise – and this approach can be successful if it can be shown that the base figures used are a reflection of the tender allowances, and are reasonable in the circumstances.

## 3.5. Burden and standard of proof

### 3.5.1 Burden of proof

Ordinarily, the burden of proof lies with the party who makes an assertion, and hence the legal maxim 'He who asserts must prove'. Accordingly, it is not for a party defending a claim to disprove the claim, it is for the party pursuing a claim to prove it. However, if the party pursuing a claim adduces sufficient evidence to raise a presumption that what is claimed is true, the burden of proof will pass to the other party. It is then for that party to adduce sufficient evidence to rebut the presumption. Therefore, any claim document should be prepared with the above basic principle in mind.

### 3.5.2 Standard of proof – what needs to be provided?

The standard of proof required in civil proceedings, is referred to as the 'balance of probabilities' principle, whereby the court or tribunal makes its decision on the basis that something is more likely to have occurred than not.

Therefore, if a set of scales is imagined as being the 'balance of probabilities', in some cases there only needs to be 51% of probability on one side of the scales to enable that side of the scales to outweigh the other side of the scales. Thus it is necessary for the party with the 'burden of proof' at any particular time to push that side of the scales down to (at least) the 51% level.

This is to be contrasted with the standard of proof in criminal proceedings, which is set at a much higher level of 'beyond reasonable doubt'. When preparing a claim (in addition to the burden-of-proof principle outlined above) the required standard of proof needs to be recognised, particularly as, in certain situations, the party receiving the claim (i.e. before proceedings are commenced) may appear to expect a standard of proof more akin to 'beyond reasonable doubt' rather than on the 'balance of probabilities'.

Another point to note is that, when assessing probabilities, the court or tribunal will have in mind, to whatever extent is appropriate in the particular case, that the more serious the allegation the less likely it is that the event occurred and, therefore, the stronger should be the evidence before the court concludes that the allegation is established on the balance of probabilities. This does not mean that different standards of proof are required where different assertions are made, but it does mean that the inherent probability or improbability of an event is itself a matter to be taken into account when weighing the probabilities and deciding whether, on balance, the event occurred. The more improbable the event, the stronger must be the evidence that it did occur before, on the balance of probability, its occurrence will be established (Lord Nicholls in *Re H (Minors)* [1996] AC 563).

It should be noted that 'more likely than not' is not necessarily a hard test to pass; and facts can often be proved to this standard merely by circumstantial evidence, or by one person's word being preferred over another's, although contemporaneous written records are always more influential. In all circumstances, some admissible evidence must be produced to support everything claimed.

## 3.6. Proof needed

Following on from the above, the following initial points need to be proved in a loss and expense claim document

- that an event actually occurred
- that the event was one expressly catered for within the contract
- that the notices required under the contract had been given
- what the effect was, in financial terms, of the specified event.

In respect of the above list, what then needs to be shown is the connection between the *cause* and the *effect*, i.e. how the last item in the above list (the financial effect) resulted directly from the first item in the list (the event that occurred). This is the part of the process that normally causes most difficulty. The connection between cause and effect is a matter of evidence and it is the part of a claim for which records are absolutely crucial. Indeed, this is when the quality and extent of records and documentation comes into its own.

With some claims such a connection is relatively easy to identify, while with others it is rather more problematic. It is because of the difficulty of establishing the required nexus of cause and effect that it is not uncommon for global claims to be produced. Loss and expense claims are considered in more detail in Chapter 5, while global claims are looked at below.

## 3.7. Global claims

A global claim is a claim for financial loss that arises from various different events. However, the requirement to link cause and effect is largely absent (i.e. individual sums of money are not claimed for each individual event), and instead a single global sum is claimed in respect of the alleged cumulative effect of all events.

Sometimes, claimants will base their claims in terms of quantum based on the difference between estimated cost and actual cost, normally with an adjustment for any recovery made within the variation account. A simple example is as follows:

| | |
|---|---|
| Total costs incurred by the contractor | £1 000 000 |
| Less the contract order value | £600 000 |
| Less recovery from variations | £200 000 |
| Claim for loss and/or expense (i.e. the shortfall) | £200 000 |

The basis of the global claim being advanced is that the total extra cost incurred (i.e. £200 000, as calculated above), has resulted from numerous events the consequences of which had such a complex interaction that it is impossible or impractical to disentangle them to show cause and effect.

Unsurprisingly, global claims have been the subject of much controversy over the years, and have been considered in many court cases. They remain a difficult and developing area of law. The general rules that appear to be being developed in respect of global claims are as follows.

- In the normal case, individual causal links must be demonstrated between each of the events for which the employer is responsible and particular items of loss and expense.
- In circumstances where it is impossible to separate the specific loss and expense caused by a number of different events which are the responsibility of the employer, these can be pleaded as producing a cumulative effect. In these circumstances, it is not necessary to break down each event and isolate the loss caused by each.
- Where, however, a significant cause of the loss and expense is a matter for which the employer is not liable, the global claim must fail.
- Where it is shown that some of these events (albeit not contributing a 'significant' amount in causal terms) are not actually the responsibility of the employer, the global claim should not necessarily fail, as it may be possible for the judge, or arbitrator, to apportion the loss between the causes for which the employer is responsible and other causes.
- It has been acknowledged that the above could lead to 'rough and ready' results.
- When pleading the claim, the particular events and heads of loss should be set out in reasonable detail. There will, however, usually be no need in commercial cases to do more than simply to plead the proposition that the particular events caused the relevant heads of loss. Causation is largely 'a matter of inference' and is frequently based on experts' reports. The consideration of the global claim should wait until this evidence is before the tribunal.

Generally, the best practice to follow in preparing a claim involving multiple variations, delay, disruption and extra cost is to, insofar as is possible, strictly follow the guidelines laid down by the courts and to present a cause and effect claim where possible, and a global claim only in respect of any element of the claim where it is simply not possible or practical to do otherwise.

While it is the case that, in the UK, global claims are very much the exception rather than the rule, in the emerging and rapidly growing international markets, global claims are more often accepted as a means of reaching a reasonably quick and fair valuation of a claim. This more liberal approach towards global claims is supported by some earlier court cases that provided justification for the use of a global claim in certain circumstances. In *J Crosby and Sons Ltd* v. *Portland Urban and District Council* (1977) 5 BLR 121, the court supported an arbitrator who had said:

> The delay and disorganisation which ultimately resulted was cumulative and attributable to the combined effect of all matters. It is therefore impracticable, if not impossible, to assess the additional expense caused by delay and disorganisation due to any one of these matters in isolation from the other matters.

And in *London Borough of Merton* v. *Stanley Hugh Leach* (1985) 32 BLR 51, Vinelott J said:

> The loss or expense attributable to each head of claim cannot in reality be separated.

Although the Privy Council in *Wharf Properties Limited and Another* v. *Eric Cumine Associates and Others* (1991) 52 BLR 1 struck out a global claim in a case where the claimant had made no attempt to link cause with effect, the comments of judges in a variety of cases since then have suggested that the courts accept that there are certain situations in which global claims may be accepted.

In *Bernhard's Rugby Landscapes Ltd* v. *Stockley Park Consortium Ltd* (QBD 1997) 82 BLR 81, the court considered all the major cases concerning global claims (at that time), and as a result produced a summary of the position (at that time). That position was that

- while a court will approach a global claim or a total cost claim with caution, such claims are not necessarily bad, and in some circumstances it may be the only way that a claimant can establish his loss
- a global claim is permissible where it is impractical to disentangle that part of the loss attributable to each head of claim, and the situation has not been brought about by delay or other conduct on the part of the defendant
- the power of the court to strike out is very limited, and should be used only where a claim is so evidently untenable that it would be a waste of resources for this to be demonstrated only after a trial
- the fundamental concern of the court is that the dispute between the parties should be determined expeditiously and economically, and above all fairly.

Even the Scottish Inner House case of *Laing Management (Scotland) Limited* v. *John Doyle Construction Limited* 11 June 2004 (Inner House, Court of Session, Scotland), which many people saw as serving the 'death knell' on global claims in the UK, actually supported the use of global claims in certain circumstances. In fact, the debate in that case with regard to the use of global claims proceeded on the 'common ground that such a claim [i.e. a global claim] could in principle be made'. In respect of causation, Lord MacLean said:

> . . . it is frequently possible to say that an item of loss has been caused by a particular event notwithstanding that other events played a part in its occurrence. In such cases, if an event or events for which the employer is responsible can be described as the dominant cause of an item of loss, that will be sufficient to establish liability, notwithstanding the existence of other causes that are to some degree at least concurrent.

He went on to say:

> ... even if it cannot be said that events for which the employer is responsible are the dominant cause of the loss, it may be possible to apportion the loss for which the employer is responsible and other causes. In such a case it is obviously necessary that the event or events for which the employer is responsible should be a material cause of the loss. Provided that condition is met, however, we are of the opinion that apportionment of loss between the different causes is possible in an appropriate case.
>
> ... Whether it is possible will clearly depend on the assessment made by the judge or arbiter, who must of course approach it on a wholly objective basis. It may be said that such an approach produces a somewhat rough and ready result ... the alternative to such an approach is the strict view that, if a contractor sustains a loss caused partly by events for which the employer is responsible and partly by other events, he cannot recover anything because he cannot demonstrate the whole of the loss is the responsibility of the employer. That would deny him a remedy even if the conduct of the employer or architect is plainly culpable, ...
>
> It seems to us that in such cases the contractor should be able to recover for part of his loss and expense, and we are not persuaded that the practical difficulties of carrying out the exercise should prevent him from doing so.

Reference was also made to several cases decided in the USA, which accepted a variant of the global claim approach, and this approach, often referred to as the 'modified total cost method', has been given approval in a recent California Court of Appeal case (*Dillingham-Ray Wilson* v. *City of Los Angeles* (2010) 182 Cal. App. 4th 1396). The modified total cost method is a variant of the 'total cost method', a methodology that compares the bid price of the project work with the total cost of the project work as performed, attributing the entire cost increase to the acts or omissions of the employer. Although advantageous to contractors, courts (as well as arbitrators and administrative boards) consistently have declared the method imprecise, and frequently have rejected it. However, the modified total cost method addresses the shortfalls of the total cost approach by segregating the aspects of the work affected by the project owner's shortcomings. Furthermore, the modified approach does not blindly accept the contractor's pre-bid cost estimates or the contractor's as-performed job cost records.

Therefore, against the above background, it may be that the use of global claims has been prematurely written off, and it may be that the commercial pressures of the developing and rapidly growing regions of the world will make the use of global claims (or at least a variant of them) more acceptable within the construction industry generally.

## Cases

*Bernhard's Rugby Landscapes Ltd* v. *Stockley Park Consortium Ltd* (QBD 1997) 82 BLR 81

*Bramall and Ogden Limited* v. *Sheffield City Council* (1985) 1 Con LR 30

*Commissioner of Public Works* v. *Hills* [1906] AC 368

*Cory & Son* v. *Wingate* (1980) 17 BLR 104, CA

*Dillingham-Ray Wilson* v. *City of Los Angeles* (2010) 182 Cal. App. 4th 1396

*Dodd Properties* v. *Canterbury City Council* [1980] 1 WLR 433, CA

*Dunlop Pneumatic Tyre Co.* v. *New Garage & Motor Co. Ltd* [1915] Ac 79

*Gaymark Investments Pty Limited* v. *Walter Construction Group Limited* [1999] NTSC 143

*Hadley* v. *Baxendale* (1854) 9 Exch 341

*J Crosby and Sons Ltd* v. *Portland Urban and District Council* (1977) 5 BLR 121
*Laing Management (Scotland) Limited* v. *John Doyle Construction Limited*: 11 June 2004 (Inner House, Court of Session, Scotland)
*London Borough of Merton* v. *Stanley Hugh Leach* (1985) 32 BLR 51
*Miliangos* v. *George Frank* [1976] AC 443, HL
*Peak Construction (Liverpool) Ltd* v. *McKinney Foundations Ltd* (1970) 1 BLR 111
*Philips Hong Kong* v. *A-G of Hong Kong* [1993] 61 BLR 41
*Re H (Minors)* [1996] AC 563
*Regional Construction Sdn. Bhd.* v. *Chung Syn Kheng Electrical Co. Bhd.* [1987] 2 MLJ 763
*Robinson* v. *Harman* (1848) 1 Ex 850
*Robophone Facilities Ltd* v. *Blank* [1966] 1 WLR 14
*Ruxley* v. *Forsyth* [1995] 3 WLR 118
*Smith New Court* v. *Scrimgeour Vickers* [1997] AC 254, HL
*Stanley Hugh Leach* v. *Merton* (1985) 32 BLR 51
*Wharf Properties Limited and Another* v. *Eric Cumine Associates and Others* (1991) 52 BLR 1

**Delay and Disruption Claims in Construction**
ISBN: 978-0-7277-5705-0


doi: 10.1680/ddcc.57050.025

# Chapter 4
# Delay and disruption

## 4.1. Introduction

Construction and engineering projects are subject to considerable risks and uncertainties including, among other things, weather conditions, ground conditions, the availability of labour, materials and plant, and changes required by the employer. Such risks and uncertainties can cause delay and/or disruption to project programmes, and delay the completion date of the project.

A *delay* event relates to delays to the overall project completion time, whereas a *disruption* event relates to changes to the planned conditions or timing of individual items of work, which may or may not affect the overall project completion time. Some events may be both delay and disruption events.

Delay and disruption events can be categorised as one of two major types: 'excusable' and 'non-excusable' events.

### 4.1.1 Excusable events

An excusable event is one for which the contractor is excused under the contract from meeting a contractual completion date or some other contractual obligation and for which he can therefore receive an extension of time and/or financial compensation (e.g. loss and expense). In general, excusable events include such things as unforeseen design problems, variations and change orders, and changes to or imposition of site restrictions. In respect of delays, this type of delay event is also commonly referred to as a 'non-culpable delay event'.

### 4.1.2 Non-excusable events

An example of a non-excusable event is time lost as a direct result of the contractor's actions or inactions. In this case, the contractor is entitled neither to an extension of time nor to financial compensation from the owner. Moreover, the contractor will be responsible for the possible impact his performance has on other involved parties. If the contract includes a liquidated damage clause, then in the case of a non-excusable delay, the owner could recover liquidated damages from the contractor. Generally, non-excusable events include the contractor's failure to perform work within the agreed time period, poor work performance and resource availability problems. In respect of delays, this type of delay event is also commonly referred to as a 'culpable delay'.

Participants in the construction and civil engineering industries often make the mistake of believing that time and money are directly linked, in that they assume that financial claim recovery can be secured *only* if an extension of time has been granted and, conversely, that it will *certainly* be due if an extension of time has been given. In reality, neither of these propositions is, strictly speaking, correct. In fact, additional monies may be recovered even if an extension of time is not applicable, or, alternatively, a contractor may receive a comprehensive extension of time and still not be entitled to additional financial recovery.

In respect of the former case, and by way of an example, it could be that scaffolding is required to remain in place for a longer period of time than originally planned on a work face that has suffered a delay, thereby causing the contractor to incur additional scaffolding hire costs which he would seek to recover as loss and expense. However, because that particular work face was not critical to the completion date of the project as a whole, no extension of time would accrue. In this case then, additional monies may be recovered, even if an extension of time is not applicable.

On the other hand, and by way of an alternative example, on a satellite project with no site set-up at all, the entire works on site could be postponed for a period of time (which would normally have the direct effect of delaying the completion date, and for which the contractor would be entitled to an extension of time) but no prolongation costs would be incurred by the contractor because there were no on-site set-up costs during the period of suspension. Therefore, in this case, a contractor may receive a comprehensive extension of time but not be entitled to loss and expense.

From the above examples, it is clear that there may be no direct link between extensions of time and loss and expense. Some delay events may create an entitlement to an extension of time but no entitlement to loss and expense, and some disruption events may create an entitlement to loss and expense but no entitlement to an extension of time. Because of this, it is always advisable that notices of delay and notices of loss and expense are kept separate from each other.

For contractors, delay and/or disruption events can result in: the imposition of liquidated or actual damages; in increased labour, material and equipment costs; in extended site and head office overheads; and in costs associated with loss of productivity. For employers, delay and/or disruption events can result in a loss of profit, revenue opportunity costs and increased consultants fees.

As the costs for both contractors and employers resulting from the delay and/or disruption events can be extensive, the liability for such events is frequently a subject of much contention between the parties.

## 4.2. The prevention principle and time at large

A contractor normally has an obligation to complete a project in a certain time period or by a certain completion date. At common law, if an employer prevents a contractor from completing his work by the contract completion date, the contractor's obligation is no longer to complete the works by the completion date as set out in the contract, but has an obligation is to complete the works within a reasonable time. This rule is normally referred to as the 'prevention principle'.

The essence of the prevention principle is that the employer cannot hold the contractor to a specified completion date if the employer has, by an act or omission, prevented the contractor from completing the works by that date. In such a case, time becomes 'at large' and the obligation to complete by the date specified in the contract is replaced by an implied obligation to complete within a reasonable time.

The prevention principle only applies where the contract does not contain a proper mechanism for allowing the employer to grant an extension of time for delays that are caused by the employer or by anyone or anything that the employer is responsible for.

The leading authority on this point is the decision in *Peak Construction (Liverpool) Ltd* v. *McKinney Foundations Ltd* (1970) 1 BLR 111. In that case, Lord Justice Salmon commented:

> I cannot see how, in the ordinary course, the employer can insist upon compliance with a condition if it is partly its own fault that it cannot be fulfilled.

As noted in Chapter 3, the prevention principle was also applied in the Australian case of *Gaymark Investments Pty Limited* v. *Walter Construction Group Limited* [1999] NTSC 143. In that case, the contractor (Walter) was delayed in completing the work, by causes for which the employer (Gaymark) was responsible and which constituted acts of prevention by Gaymark. Under the provisions of the contract, Gaymark was not able to grant an extension of time to Walter, nor was it able to set a revised completion date, and, as a result of this, time became at large and Walter was simply obliged to complete the work within a reasonable time.

In *Multiplex Constructions (UK) Ltd* v. *Honeywell Control Systems Ltd* (TCC 6 March 2007), Honeywell (a subcontractor) contended that the project was so mismanaged by Multiplex (the main contractor) that it was no longer bound by the terms of its subcontract to complete its works by a fixed date. Honeywell argued that time had become at large and that its obligation to complete its subcontract works within a specified time had fallen away, and had been replaced by an obligation to complete the works within a reasonable time. Honeywell founded its argument on the prevention principle. Multiplex commenced a pre-emptive action in the court to secure a decision that the contract mechanisms for fixing the date for completion remained operative. In those proceedings, Honeywell argued that Multiplex had made the extension of time machinery inoperable by its failure to provide adequate programming data to Honeywell, which, in turn, meant that Honeywell could not effectively comply with the subcontract requirement for notices and particulars of delay to be given, which in the subcontract was made a pre-condition of Honeywell's right to receive extensions of time. In court, and on a matter of construction, HH Justice Jackson found that the sweep-up provisions of clause 11 of the subcontract in question enabled an extension of time to be given by the contractor for an act of prevention, irrespective of any notices and particulars of delay being provided by the subcontractor, and, consequently, time was not at large on the grounds that had been asserted by Honeywell. HH Justice Jackson also found that the extension of time machinery had been properly operated and had not broken down in any event. Therefore it was found that, even though it is well-established law that a party cannot insist on the performance of an obligation, which it has prevented the other party from performing, in this case the subcontract permitted Honeywell to obtain an extension of time, despite the possible procedural problems created by its inability to provide delay notices, and thus, and despite Honeywell's assertion, time was not to be considered as being 'at large'.

However, despite the findings in this case, there are situations where 'time at large' is applicable. In such cases, 'time at large' does not mean that the contractor does not have an obligation to complete the works, but it does mean that the contractor is not necessarily bound by the original agreed contract time period, but rather is required to complete the works in a reasonable time.

Of course, this opens up the question of what is a 'reasonable time'. A review of the court cases shows that what is considered to be a reasonable time depends largely on both what was anticipated at the outset of the project, including perhaps the anticipated level of resources, and also on those factors that may have caused delay during the project and over which the contractor had no control. Therefore, reasonable time is primarily a question of fact and it must depend on all the circumstances that might be expected to affect the progress of the works.

The concept of reasonable time was considered by J. Diplock in *Neodox Limited* v. *The Borough of Swinton and Pendlebury* [1958] 5 BLR 38. That case involved the question of whether an engineer acting for the defendant had failed to issue instructions to the contractors within a reasonable time. J. Diplock ruled that what was a reasonable time did not depend solely on the convenience and financial interests of the contractor. He observed that, while it may appear to the contractor that it is in his interest 'to have every detail cut and dried on the day the contract is signed', such a state of affairs could not have been contemplated at the time when the contract was made. J. Diplock then proceeded to hold that what was a reasonable time was a question of fact to be determined with reference to all the circumstances of the case.

In *British Steel Corporation* v. *Cleveland Bridge & Eng Co.* [1981] 24 BLR 100, it was decided that what constitutes a reasonable time has to be considered in relation to the circumstances that existed at the time when the contract obligations were performed, but excluding circumstances that were under the control of the contractor. Lord Goff applied these principles by first considering what a reasonable time for performance would be in ordinary circumstances and then considering to what extent the time for performance of the contractor should in fact be extended by extraordinary circumstances that were outside of his control. Lord Goff noted that whether a reasonable time has been taken to do the works cannot be decided in advance, but can only be decided after the work has been completed.

## 4.3. The purpose of extension of time clauses

As noted above, the prevention principle applies only where the contract does not contain a proper mechanism for allowing the employer to grant an extension of time to the contractor for delays that are caused by the employer or by anyone or anything that the employer is responsible for.

Conversely, the prevention principle does not apply in the case where there is an express provision in the contract for the employer to set a new completion date in the case of acts of prevention by the employer. Where such an express provision does exist in the contract, the employer can grant an extension of time for acts of prevention, and, because of that, the employer can preserve his rights to claim liquidated damages from the contractor if the contractor fails to complete the works by the (extended) completion date.

Consequently, and in order to avoid the effect of the prevention principle, most construction contracts include express provisions for extensions of time to be granted in respect of acts of prevention by the employer or by anyone or anything that the employer is responsible for. The existence and proper operation of such an extension of time provision can therefore be seen to be more for the benefit of the employer than the contractor, as, without such an express provision existing, the employer would be unable to exercise his rights to recover liquidated damages if the contractor failed to meet the completion date because of acts of prevention as described, and time would, in effect, become at large. On the other hand, and in view of the fact that a contractor's entitlement to an extension of time is a separate issue to his entitlement to loss and expense, the only real benefit to the contractor of obtaining an extension of time is a relief of its liability for the payment of liquidated damages for delay.

It is important to note that the extension of time provision referred to above only needs to be able to operate in the case of acts of prevention that the employer has control over, and therefore the inability of the contractor to obtain an extension of time as a result of delays caused by other events (e.g. inclement weather) would not result in time becoming at large should the contractor suffer delays as a result of that other event. In *Peak Construction (Liverpool) Ltd* v. *McKinney Foundations Ltd* (1970) 1 BLR 111, the works were suspended due to defective piles for which the contractor was responsible, but the employer caused further delay for which there was no mechanism in the contract for extending time. The main contract allowed the architect to certify extensions of time for additions to the works, strikes, *force majeure* or any other unavoidable circumstances. These provisions did not permit an extension of time to be granted for a failure by the employer to promptly authorise and instruct the investigations and remedial works. In that case it was held that, to entitle the employer to grant extra time for his own defaults, the extension of time clause must provide, expressly or by necessary inference, for an extension of time on account of a fault or breach on the part of the employer. Many standard form contracts now contain a catch-all clause that allows the employer to grant an extension of time in respect of any acts of default by the employer, which is designed to avoid the prevention principle from being applied.

## 4.4. Project programming and float

In the absence of an express term, there may be an implied term in every construction contract that the contractor shall commence the works within a reasonable time of making the contract and proceed with the works at a reasonable rate. The completion date of the works given within a contract is critical as, not only does it have a direct bearing on the question of whether the employer can levy liquidated damages on the contractor, but it also usually marks the transfer of certain risks or the crystallisation of certain rights between the contractor and the employer.

There is no implied term that a contractor is required to execute the works with any particular due diligence and expedition, but only with such diligence and expedition as would reasonably be required in order to meet the key dates and the completion date in the contract. Accordingly, the contractor has the freedom to plan his work as he sees fit within the specified time constraints and is not obliged to proceed at a particular work rate. He must ensure, however, that progress is not so slow that he can be said to be deliberately putting himself in a position where he cannot complete on time, which could, in an extreme case, amount to an anticipatory repudiatory breach of contract.

Clearly, commencing, proceeding and completing the works are three of the major components in any construction project where time factors are essential in interconnecting the different variables affecting the process of executing construction in a timely manner. The relationship between these factors is complex and creates a web of interactive activities that need to be identified before the start of the works and need to be updated regularly throughout the project. This is achieved by the use of project programming.

Project programming consists of updating current and target schedules for existing projects, and developing breakdown structures, milestones, target schedules and cost-loaded schedules for new projects. Most standard forms of construction contract contain provisions for the contractor to submit to the employer a programme showing the manner in which he intends to carry out and complete the works. That requirement may range from a simple request for the submission of a master programme, without prescription as to its form or content, to the requirement that a very detailed programme is provided in a stipulated format, and with a specified content.

In whatever form it is submitted, the programme is a crucial document for the effective management of most construction projects, as it provides a tool by which actual job progress is monitored against a plan, thus enabling an early alert of actual and potential delays which could adversely affect the project completion date. Therefore, to reduce the number of disputes relating to delay, the contractor should prepare a programme showing the manner and sequence in which he plans to carry out the works. The programme should be updated to record both actual progress and any extensions of time granted.

As the complexity of projects and the requirements for scheduling have increased, so have the opportunities for delay to the various activities that have been scheduled and are necessary for the completion of the project. The resolution of disputes regarding the effect of delay and/or disruption events on large construction and engineering projects increasingly involves the use of delay analysis techniques to assist in the identification of the causes of critical delay to a project and to assist in the computation of claims for lost productivity.

In terms of the programme used during the contract and post-contract period, the analysis will not necessarily establish the impact and extent of delay unless more sophisticated forms of analysis are adopted and are used on an ongoing basis throughout the project, such that the programme becomes literally a working and living document. This is because if delay allegations are to be shown effectively by the contractor and considered properly by the architect or the engineer, it will be found that, in most situations, a simple bar chart will not suffice and some better means of indicating quantity output or physical progress, as well as the passage of time, are essential.

The price of failing to establish a proper programme is well illustrated by the case of *Balfour Beatty Construction Limited* v. *The Mayor and Burgess of the London Borough of Lambeth* (2002) 1 BLR 288. In that case, and in reaching his decision that Balfour Beatty was unsuccessful in its action, HHJ Humphrey Lloyd observed that:

> Despite the fact that the dispute concerned a multi-million pound refurbishment contract, no attempt was made to provide any critical path. It seems that Balfour Beatty had not prepared or maintained a proper programme during the execution of the works. By now, one would have thought that it was well understood that, on a contract of this kind, in order to attack, on the facts, a clause 24 certificate for non-completion (or an extension of time determined under clause 25), the foundation must be the original programme (if capable of justification and substantiation to show its validity and reliability as a contractual starting point) and its success will similarly depend on the soundness of its revisions on the occurrence of every event.

As to the ownership of float in a programme, the parties should expressly address the issue in their contract. Float is the time available for an activity or a series of activities in addition to the actual time required to carry out that activity or series of activities. There are two main types of float: an activity float, sometimes referred to as 'free float', which is the time available to an individual activity without it affecting a succeeding activity; and terminal float, sometimes referred to as 'project float', which is the free time shown after the completion date shown on a programme up to the contract completion date. The ambiguous interpretation of total float ownership can be clarified by improving contract language with regard to scheduling specifications in the area of total float management. The proposed concept for managing total float should involve pre-allocating a set amount of total float on the same non-critical path of activities to the two contractual parties – the employer and the contractor.

Without this consideration of the question of total float, court cases have found that there is no implied term in building contracts that the employer should perform the contract so as to enable the contractor to complete the works in accordance with a programme showing a completion date earlier than the contractual completion date (i.e. a programme with a terminal float period). In *Glenlion Construction Ltd* v. *Guinness*

*Trust* (1987) 39 BLR 89 QBD, Glenlion were contractors under a JCT form of contract for the construction of a residential development for Guinness in Bromley, Kent. The issue that arose was whether Glenlion was entitled to complete the works in line with its submitted programme, which showed a completion date before the contract completion date. In court it was held that it was self-evident from clause 21 of the JCT contract that Glenlion was entitled to complete before the date of completion, and this was so whether or not Glenlion produced a programme with an earlier date and whether or not it was required to produce a programme at all. However, crucially, while it was found that Glenlion was entitled to complete the works before the contract completion date, it was also held that Guinness was not required to actively cooperate to enable the earlier date to be achieved, but was only required not to hinder earlier completion. It must be noted that the judgment in the Glenlion case was founded on the terms of the JCT contract, and a different conclusion would almost certainly have been reached had the contract been formed using a different standard form (e.g. NEC 3).

As is clear from the above, the *float ownership* concept is fundamental to the analysis of project delay. Both the employer and the contractor want access to the float in a programme because it affords them more flexibility in their decision-making and use of resources. However, many contracts do not address this important topic. As a result, neither the employer nor the contractor has a contractual right to use the float. The now generally accepted, and sometimes disputed, answer is that *the project owns the float*. Under this interpretation, a party is permitted to delay an activity with positive total float, provided that the delay duration does not exceed the total float calculation for that activity, and their use of the positive total float occurred prior to anyone else's. The float ownership concept becomes much more complex when the project is late, and the total float calculation becomes, in effect, negative.

The NEC 3 form of contract clearly states that there is no reason why the contractor cannot show on his programme a date for planned completion earlier than the contract completion date, thus including some terminal float in its programme. NEC 3 core clause 63.3 provides that: 'A delay to the completion date is assessed as the length of time that, due to the compensation event, planned completion is later than planned completion as shown on the accepted programme.' Therefore, as stated in the guidance notes to that contract, any terminal float (but not activity float) resulting from an early planned completion date is preserved. The period of delay to the planned programme is then added to the completion date to determine the revised completion date from which delay damages will be applicable.

The above principle flows from the fact that, under NEC 3, the programme is a much more important tool for the administration of progress and delay. Core clause 31.2 sets out the requirements of each programme that the contractor submits, including the requirement to show the completion date (as stated in the contract data or as extended by the contract), planned completion (which may be a different date) and any provisions for float. The intention, in accordance with core clause 32.2, is that the programme is revised monthly or when the contractor chooses or the project manager instructs. The programme is required to reflect the state of affairs at that time (i.e. it is to represent the as-built status of the works), and is to include the effect of any compensation events for which an allowance for delay has been determined, as well as other delays to date.

## 4.5. Critical path method

Linked to the concept of 'float' is one of the primary, if not the single most important, methods for project programming – the critical path method scheduling technique. The provision of critical paths is widely accepted and used in the construction industry because they serve not only as a tool for planning projects, but also as a means of minimising time-related claims, to justify actual claims, and to assist in negotiating timely solutions of both in and out of court settlements.

The critical path method is a tool that demonstrates the shortest possible time to completion at any stage by breaking down the interrelationship of the discrete elements of a programme that comprise the activities to be undertaken. A well-constructed programme using the critical path method allows the parties to identify which parts of the project, called 'activities', are critical to the completion date. The longest path of the resulting schedule of linked activities is called the 'critical path'. This path consists of activities that, if delayed, will extend the project beyond the completion date identified on the programme. In addition to the critical path, there are various other side paths called 'non-critical paths'. If affected by improper scheduling or performance delays, these non-critical paths could themselves (if delayed for a long enough period) become critical and thus alter the original critical path. A critical path method is designed to advise involved parties about the relative importance of performing certain activities within the project completion parameters. It indicates to participants whether their work is critical or non-critical, and whether it has any float associated with its performance.

The advantages of using a critical path method can be summarised as follows.

- A critical path requires a detailed analysis of the project, and therefore the programmer will have a better understanding of the project. This requirement minimises the possibility of erroneous or misleading schedules.
- Critical paths are well established and easy to understand, the techniques for drawing and calculating project durations having been developed from advanced technologies and using standard quantitative software.
- The critical path method can be used to determine the length of delays or additional times needed as a result of unexpected events occurring or changes demanded during the construction process.

Courts have accepted the use of the critical path method to prove delay, to identify the causes of such delays and inefficiencies, and to assign responsibility for them. J. Dyson, in *Henry Boot Construction (UK) Ltd* v. *Malmaison Hotel (Manchester) Ltd* (2000) CILL 1572 TCC, confirmed the importance of the critical path method in establishing causation, and also confirmed that a delay is only relevant when it falls on the critical path. This principle also received support in *The Royal Brompton Hospital NHS Trust*

v. *Hammond & Others* (2001) 76 Con LR 148, QBD (TCC), where J. Seymour, *obiter dicta*, confirmed that in determining a fair and reasonable extension of time as a consequence of a delay event, an examination of the actual critical path of the contractor's programme should be carried out to establish that the delay event affected, or was likely to affect, the completion of the works.

US construction case law dominates the references to the critical path method, and US judges show a willingness not only to understand the mechanism of a critical path method but also to prefer this approach to other programming methods, such that the critical path method is commonly used to prove cause and effect in delay-related disputes.

A key point with the critical path is that it should not be seen as being static, but must be understood to be a dynamic path that can change depending on the circumstances of and the delays encountered on the project in question.

In *Natkin & Co.* v. *George A. Fuller Co.* 347 F. Supp. 17 (W.D. Mo. 1972), one of the findings adopted by the judge in order to reach his decision was that 'the critical path plan may become obsolete unless it is kept current'. The judge emphasised that the usefulness of a critical path method as a barometer for measuring time extensions and delay damages is necessarily circumscribed by the extent to which it is employed in an accurate and consistent manner to comport with the events actually occurring on the job. Furthermore, updating the critical path during the life of the project is incremental, and if such updating is not done the schedule can be made redundant. This point was emphasised further in *Fortec Constructors* v. *United States* 8 Cl. Ct. 490 (1985), where it was found that 'if the critical path method is to be used to evaluate delay on the project, it must be kept current and must reflect delays as they occur'.

### 4.5.1 Critical path weaknesses

When carrying out retrospective delay analysis using critical path analysis, large programmes with hundreds and thousands of activities can produce unreliable results. This is because the analyst has made hundreds of assumptions in preparing the programme and, when considering the impact of an event, he would likely make many adjustments to a programme if faced with a potential delay, especially if the delay is on the critical path.

It is important to recognise that it is easy (often inadvertently) to manipulate a critical path in order to derive the required end result. Programmers can alter the sequence and the scheduling mechanism to create preferred results for the party construing it. For example, if a programmer wishes to make a certain section of the work critical, he achieves this by fixing durations of activities or logic links between activities. Equally, if variations in part of the works have been issued, it is possible to make this element of the programme critical. There is also a tendency for those involved in preparing the critical path method to get lost in the analysis rather than focusing on the important task of establishing the entitlement arising from the events to be modelled.

Courts have, on occasion, had difficulty understanding and accepting the logic behind a critical path method, together with other programming issues such as the processes commonly used to prove cause and effect in delay-related disputes. In certain cases, judges have expressed their preference for analysis based on factual evidence, sound practical experience and common sense, despite the fact that such analysis might not be based on critical path analysis, and have jettisoned the critical path approach entirely where it has been based on flawed as-built critical path analysis.

A case in point is the *City Inn* v. *Shepherd Construction* case which concerned the construction of a hotel under an amended JCT form. Matters in dispute included the pursuer, City Inn, seeking a declaration that the defendant was not entitled to the contended 11-week time extension and not even to the 4-week time extension that had been granted by the architect. In reaching its decision, the Inner House of the Scottish courts (on appeal from the Outer House of the Scottish courts) found that a common sense approach must be used when deciding whether or not the cause of a delay is an excusable event, and added that, although a critical path may be of assistance in determining causation, it is not indispensable. The court found that, above all, an architect (for example) should approach the issue of an extension of time entitlement in a fair and reasonable manner, should apply common sense and should not be overly restricted by a theoretical critical path analysis (particularly one which, as in this case, was found to be flawed or incomplete).

The above cases suggest that, although critical path method techniques are recognised as appropriate for delay analysis, it is very important for contractors and employers or their agents to employ techniques that consider what actually happened on site based on factual evidence. Theoretical delays calculated without taking into account actual project records are unlikely to succeed.

Having said this, it must be noted that an architect (for example) cannot move too far away from carrying out some form of detailed analysis of the delay. This is because in *John Barker Construction Ltd* v. *London Portman Hotel Ltd* (1996) 83 BLR 31 it was established that, in exercising his duty, an architect must undertake a logical analysis in a methodological way of the impact of the excusable events on the contractor programme, and that the application of an impressionistic rather than a calculated and rational assessment is not sufficient.

## 4.6. Events that cause delay

The most important, and indeed difficult, task in terms of determining delay is to establish the nexus of cause and effect; the mere happening of an event confers no entitlement to an extension to the contract period.

The test is whether the event in question actually caused a delay to the completion date (or to a section thereof). In considering this matter, it is generally actual progress that must be delayed and the originally planned/programmed progress is not generally relevant (*Glenlion Construction Ltd* v. *Guinness Trust* (1987) 39 BLR 89 QBD). However, under the NEC 3 contract, it is delay to the accepted programme (which is itself a dynamic document) that is considered.

In all cases, what need to be considered are the actual facts on site to determine whether a particular event delayed the completion date. What has to be established in determining an extension of time entitlement can be briefly described as follows:

- That an event occurred at all.
- That the event occurred in the manner asserted.
- That the event comes within a category providing an express entitlement to an extension of time within the contract.
- That the required notices were given regarding the cause of delay and identifying the clause in the contract, which means that the specified cause entitled the contractor to an extension of time. Where the contract provides for specific procedures to be followed as a condition precedent to the valid exercise of the power to deduct liquidated damages, for example a written application from the contractor or a certificate issued by the employer's contract administrator, such a procedure must be complied with. Whether the giving of a notice is a condition precedent is a question of construction of the particular contract.
- That the event outlined had a particular delaying effect on the completion date. Normally, the contractor is required to provide with the notice, or as soon as possible after the notice, particulars of the expected effects of the event and an estimate of the expected delay to the completion of the works beyond the completion date.
- Upon receipt of the notice and any further particulars the employer's contract administrator is required to decide whether in his opinion any of the events notified are relevant events and whether as a result of such events the works are likely to be delayed beyond the completion date.
- If the employer's contract administrator so decides, he is then required to give an extension of time to the contractor in writing such as he then estimates to be fair and reasonable. Often, the extension of time clause stipulates that the extension of time should be fair and reasonable and should apply the rules of the contract. What constitutes the basis of a fair and reasonable extension of time was considered in *John Barker Construction Ltd* v. *London Portman Hotel Ltd* (1996) 83 BLR 31, where HHJ Toulson QC set out the criteria that must apply in order to calculate a fair and reasonable extension of time, which comprises the fact that the architect (in that case) recognises the effects of constructive change, makes a logical analysis, in a methodical way, of the effect of relevant events on the contractor programme, and calculates, rather than makes an impressionistic assessment of, the time taken up by relevant events.

## 4.7. Delay analysis

Delay claims are normally demonstrated by a logical interpretation of the events. Such claims can be supported by documents, letters, instructions and witness statements, or may extend to the use of computer-aided project management tools. In essence, what is being sought is the delay that is caused by an excusable event occurring at the time when it in fact occurred, with the project being in the state that it was in at that time, and with the contractor responding to it as he did, with an allowance then being made for any extent to which the contractor has, through breaches of contract, contributed to the resulting delay. Such a contribution might have affected the state of the project at the time when the delaying event occurred, or might have affected it afterwards if, for example, the contractor failed to comply with an obligation to use his best endeavours to overcome a delay.

Mostly, proof of delay is a matter of fact, but sometimes facts alone cannot answer the question and the law is required to take a position.

There are various approaches used by programming specialists to demonstrate or assess the delaying effect of particular relevant events. The method that is actually adopted is often decided on according to factors such as proportionality (i.e. the cost of producing the delay analysis as compared to the total value or amount in dispute) and what materials are available in order to construct the programme and progress details required. Some of the more commonly used delay analysis techniques are outlined below.

### 4.7.1 Bar chart methodology

Contractors seeking to prove entitlement to time or money often use a bar chart comparing the as-planned and the as-built programme. The contractor will then claim relief for the difference between the end dates shown on the two programmes.

### 4.7.2 Retrospective critical path method

This method of analysis has revolutionised the way in which construction projects are programmed and managed, and the way in which the effect of delay can be predicted and calculated. Its main advantage is the opportunity that it provides to link cause and effect at a level of considerable detail. There are several variations on the retrospective critical path method analysis, including

- *as-planned impacted*, which adds client-caused delays into the as-planned programme
- *as-built but for*, which subtracts employer-caused delays from the as-built programme.

The two predominant areas of retrospective delay analysis are static and dynamic. Static critical path analysis is largely inferior to dynamic critical path analysis and is usually adopted when the cause is clearly identifiable, and there are no complex issues such as acceleration or unproductive work, and there is no change in the logic. However, because construction contracts are themselves dynamic, a dynamic critical path methodology is the preferred route to retrospective delay analysis. Dynamic critical path methods are classified as time-impacted delay analysis and are based on an analysis of the delaying events at the time when they occur.

### 4.7.3 Window analysis

In this approach, the construction process is seen as a series of windows during the period of performance. For each window the programme is updated to take account of delays that are at the contractor risk, any necessary logic or duration revisions by reason of mitigation, and all excusable and compensable events during the period since the last update. Thus, each programme update will incorporate all changes that affected the planned progress of the project during that period. This method is most effective when used contemporaneously, and should be updated regularly throughout the course of the project. As each window is only a segment of the contract period, the result of each window analysis must be summarised and carried forward to the next window. It is thus only when the last window, closing at actual completion, has been analysed and summarised that the accumulation of the various changes can be added together to demonstrate the effect on completion of the various events. In simple terms, what is being constructed is an as-built programme that shows the impacts of all delaying events as they occur.

For the window analysis to be effective, accurate progress information at the time of the windows must be available, otherwise the analysis cannot be properly or accurately completed. The less accurate the available programme and progress information, the more likely that results will be obtained that are clearly inaccurate, and that will need to be amended by manipulating any obvious errors in the original as-planned programme.

### 4.7.4 Snapshot analysis

In a snapshot analysis it is the occurrence of the event itself and the cessation of its operation which dictate what is analysed and at what point in the progress of the project the analysis is done. This is a simple contemporaneous approach to delay analysis which allows assessment to be made of three important aspects

- the actual state of progress at the time the delaying event occurred
- the changing nature of the critical path as a result of the events
- the effects of action taken, or which should reasonably have been taken, to minimise delays or avoid subsequently delays.

## 4.8. Concurrent delays

In certain instances, concurrent delays can occur, i.e. two delay events that have an impact on the critical path to completion occur at the same time. If the two delay events are not the liability of the contractor (or, in fact, if both delay events are the liability of the contractor) then the question of concurrent delays is not a major concern. The difficult issue in respect of concurrent delays occurs when one of the delay events is not the liability of the contractor (i.e. it is an excusable delay event), but the other event is (i.e. it is a non-excusable delay event). The question that then arises is how the delay to completion is allocated between the two concurrent delay events.

Concurrency has been, and continues to be, a highly contentious legal and technical subject in engineering and construction projects. This is largely due to the fact that resolving issues of concurrency requires the consideration of the interaction of different factors, such as the time of occurrence of the delays, their length of duration and criticality, the legal principles of causation and float ownership. The resolution of concurrency also requires the consideration of the views of the parties involved and the mitigation steps taken by them, such as reallocation of resources, incentives for acceleration procedures and delay pacing strategies.

The courts have dealt with the matter of concurrency in many different ways, and some of the more common approaches used are the American approach, the chronology of events approach, the dominant cause approach, the application of the 'but for' test, the common sense approach, the Malmaison approach and the apportionment approach. Some of the more important of these approaches are described below.

### 4.8.1 The American approach

In the American approach, which is based on US case law, the general view on concurrent delays is that the employer and the contractor are both responsible for delays to project completion, in that neither party will recover financial recompense unless and to the extent that they can segregate delay associated with each competing cause.

### 4.8.2 The application of the 'but for' test

By this test, a party seeks to lay responsibility for project delay on the other party by arguing that the delay would not have occurred 'but for' certain actions or inactions that occurred before the concurrent delay event. Although such arguments are often made, there appear to be very few reported court cases that lend support to the use of the test. However, in the case of *The Royal Brompton Hospital NHS Trust* v. *Hammond & Others* [2001] 76 Con LR 148, QBD (TCC), HHJ Seymour provided some support to the aforementioned test, when he said:

> However, if Taylor Woodrow was delayed in completing the works both by matters for which it bore the contractual risk and by relevant events, within the meaning of that term in the standard form, in light of the authorities to which I have referred, it would be entitled to extensions of time by reason of the occurrence of the relevant events notwithstanding its own defaults.

### 4.8.3 The common sense approach

The common sense approach is where the question of causation is treated by the application of the logical principles of causation. However, in the case of *H. Fairweather and Co. Ltd* v. *London Borough of Wandsworth* (1987) 39 BLR 106, the court, *obiter dicta*, considered that applying the common sense approach to concurrency was not sufficient on its own merit, and that each separate cause of delay should be assessed individually. Another weakness of this approach is the common sense criterion relied on, which could result in unfair apportionment, particularly where the competing causes are of approximate equal causative potency. Additionally, the approach may not suffice on projects that have sustained multiple overlapping changes or delays of long duration, because of all the assumptions that must be made regarding the remaining durations of activities being affected, which means that the programme becomes too hypothetical to apply.

### 4.8.4 The Malmaison approach

The Malmaison approach is named after the case of *Henry Boot Construction (UK) Ltd* v. *Malmaison Hotel (Manchester) Ltd* [1999] 70 Con LR 32. In the Malmaison approach, which is one of the most widely adopted and accepted in the industry, provided one of the causes of delay in any given concurrency situation affords grounds for extension of time under the contract, the contractor should be given an extension of time notwithstanding any default on his part. The approach sounds reasonable and just, in the sense that denying the contractor an extension of time in such circumstances could make him liable for the payment of liquidated damages even though the project would have been delayed anyway due to employer's default. In his judgment, HHJ Dyson stated:

> It is agreed that if there are two concurrent causes of delay, one of which is a relevant event, and the other is not, then the contractor is entitled to an extension of time for the period of delay caused by the relevant event notwithstanding the concurrent effect of the other event. Thus to take a simple example, if no work is possible on a site for a week not only because of exceptionally inclement weather (a relevant event), but also because the contractor has a shortage of labour (not a relevant event), and if the failure to work during that week is likely to delay the works beyond the completion date by one week, then if he considers it fair and reasonable to do so, the architect is required to grant an extension of time of one week. He cannot refuse to do so on the grounds that the delay would have occurred in any event by reason of the shortage of labour.

### 4.8.5 The apportionment approach

In the absence of there being an identifiable dominant cause, the person assessing the entitlement to an extension of time may be expected to apply the apportionment approach to the competing concurrent causes. In determining apportionment in a case of concurrent delays, the following principles must be followed.

- Before any claim for an extension of time can succeed, it must be shown that the relevant event is likely to delay or has delayed the works. Whether the relevant event actually causes delay is an issue of fact which is to be resolved by the application of principles of common sense.
- The decision-maker can decide the question of causation by the use of whatever evidence he considers appropriate. If a dominant cause can be identified in respect of the delay, effect will be given to that by leaving out of account any cause or causes that are not material.
- Where there are two causes operating to cause delay, neither of which is dominant, and only one of which is a relevant event, a contractor claim for an extension of time will not necessarily fail. Rather, it is for the decision-maker, approaching the issue in a fair and reasonable way, to apportion the delay in completion of the works as between the relevant event and the other event.

In *John Doyle Construction Limited* v. *Laing Management (Scotland) Limited* (2002) CILL 1870, Lord Young was of the opinion that it may be possible to apportion the loss between the causes for which the employer is responsible and other causes if it can be said that events for which the employer is not responsible are the dominant cause of the loss. In such a case it is obviously necessary that the event or events for which the employer is responsible should be a material cause of the loss. Provided that this condition is met, however, apportionment of loss between the different causes is possible in an appropriate case, where the causes of the loss are truly concurrent, in the sense that both operate together at the same time to produce a single consequence.

In *City Inn* v. *Shepherd Construction* (2007) CSOH 190, a Scottish case, the sitting trial judge was again Lord Young, and he adopted the approach of apportionment when it came to considering concurrent delays. He said, *obiter dicta*, that if a dominant cause could not be established between two competing causes, one a relevant event and the other a contractor default, then the architect or a tribunal could apportion the effects of these delays.

Lord Osborne, in the appeal case of *City Inn* v. *Shepherd Construction* (2010) CSIH 68, in agreeing with Lord Young, said that where there are concurrent causes it will be possible for an architect or other tribunal to apportion delay to the completion of the works between the competing causes, assuming that there is no evidence of a dominant cause.

### 4.8.6 Summary

An analysis of the numerous court cases involving concurrent delays shows that, in reality, the courts deal with each case on its own merits, and there are no hard and fast rules when it comes to matters of concurrency. It appears that the courts are concerned primarily with what has actually happened, rather than considering any sort of analysis that may be based on some speculation about what would or could have been the effect at the time when the event occurred. Judges are really looking to get to the facts in order to find out what really happened on the site and to identify the real causes of delay. It is important to recognise that there may be a distinction between what really caused delay on the project and a contractor's entitlement to an extension of time in accordance with the terms of the contract.

## 4.9. Extension of time in a period of culpable delay

There are two schools of thought, or methods, on how an extension of time should be calculated where an extension of time is granted during a period of non-excusable delay (i.e. a delay that is wholly the responsibility of the contractor).

The first method is described as the 'gross' method, and is the preferred approach of many academics and some commentators. The method propounds that, if an extension of time is granted because of an event arising during a period of culpable delay, then the extension of time must begin to run from the date the event that occurred was given effect. This means that the architect must establish a new completion date for the contract, which adding the extension of time from the date of the instruction, and thus denying the employer liquidated damages up to the new completion date. Naturally, many employers find this to be unfair, and this has traditionally been a contractor-led argument. This argument arises out of the case of *Wells* v. *Army & Navy Cooperative Society* (1902) 86 LT 764.

The second method, known as the 'net' method of calculation, was preferred by Lord Denning (in *Amalgamated Building Contractors Ltd* v. *Waltham Holy Cross UDC* (1952)). In such situations, it is argued that the contractor is only entitled to an extension of time equal to the time required to carry out the additional work. Effectively, this means that if the contractor is 6 months in delay and is delayed by one further month due to a an excusable delay event; the completion date would be extended from the original completion date by 1 month only, still leaving the contractor with 5 months of culpable delay and the threat of liquidated damages. Some contractors would consider this to be unfair, as the employer may be directly responsible for the late excusable delay event (e.g. issuing instructions for extra work). However, in *Balfour Building Ltd* v. *Chestermount Properties Ltd* (1993) 62 BLR 1 QBD, the court confirmed that the purpose of the power to grant an extension of time was to fix the period of time by which the period available for completion ought to be extended, having regard to the incidence of excusable delay events. The completion date, as adjusted, was not the date by which the contractor ought to have achieved practical completion, but was the end of the total number of working days starting from the date of possession, within which the contractor ought fairly and reasonably to have completed the works. On this footing, J. Coleman, *obiter dicta*, in the Balfour Beatty case mentioned earlier, clarified the issue, stating that, where a relevant event arose after the date for completion and during a period in which the contractor was in culpable delay, the contractor would only become entitled to a net extension of time corresponding to the specific number of days of delay occasioned by the excusable delay event (this is sometimes referred to as the 'dot-on' approach to extensions of time). In other words, the occurrence of the new delaying event would mean that the contractor was still liable for its own culpable delays. He concluded:

> Accordingly, I conclude on the second question that it would be wrong in principle to apply the 'gross' method, and that the 'net' method represents the correct approach.

## 4.10. Records

Most contractors, during the course of a contract, collect the type of information necessary to enable them to particularise events. While the documentation and recording of such information is primarily for management purposes, it also places the contractor in a good position if he needs to pursue a claim. Those who fail to collect and document information on changes and delay events will be left with few options. Attempting to collect and assemble data after the event will limit claims to retrospective accounts and analysis. It should be noted that the courts place a particularly heavy burden on contractors in terms of the maintenance and presentation of documentation in support of any claim for delay. The contractor has to maintain accurate and complete records, and should be able to establish the causal link between the client risk event and the resultant delay or disruption and/or loss and/or expense suffered. There is much truth in the old saying that the three most important constituent parts of a successful claim are 'records, records and records'!

## 4.11. Acceleration

One of the methods that contractors will employ to counteract the impact of delay and disruption is to accelerate the work on the project. This involves speeding up aspects of the project that have potential to be accelerated, either because of the nature of the task or because of the resources or capabilities of the contractor. In either case, strengths of the contractor in one area are being used to compensate for weaknesses in another. Acceleration may be achieved by a change in the deployment of resources, longer working hours or additional days of working with the same resources. In some cases, it may be achieved by simply changing the order or sequence of carrying out the work, and may therefore not incur additional cost.

Despite this latter possibility, acceleration can affect a contractor's costs in a variety of ways, such as

- additional labour and equipment costs arising from reduced efficiency of the expanded labour force and supplied equipment
- additional delivery charges for material and equipment required at the site outside of normal work hours
- costs of additional site facilities
- additional costs of advancing the date of delivery of manufactured elements
- overtime charges for operatives, engineers, staff and foremen.

Acceleration measures may be the a result of an employer choosing to reverse the impact of delays by expressly ordering acceleration in order to put the project back on schedule. If the contract provides for acceleration, payment should be based on the terms of the contract. If the contract makes no provision, the parties should agree the basis of payment before acceleration is commenced.

Alternatively, where the employer resists granting an extension of time to a contractor which the contractor is entitled to for an excusable event, or where an extension has been granted by the employer, but for a shorter period than the contractor is entitled to, a contractor may feel compelled to accelerate the works in order to overrun the completion date set by the employer, thereby avoiding exposure to liquidated damages. This is often referred to as 'constructive acceleration' and if it is hoped to recover under this head of claim the contractor must prove

- that an extension of time was requested for an excusable delay according to the contract provisions
- the employer failed to grant an adequate, or any, extension of time
- the employer made it clear that completion was required within the original contract period (or within the period that the employer had agreed to extend the completion date to)
- adequate notice had been given by the contractor to the employer advising that he was treating the employer's actions as constructive acceleration
- that additional costs had actually been incurred as a result of the constructive acceleration measures undertaken.

In *John Barker Construction Ltd* v. *London Portman Hotel Ltd* (1996) 83 BLR 31, one of the issues was whether liquidated damages could be imposed when acceleration was agreed between the parties on the sectional completion provisions of the contract. In this case delays occurred, and it was apparent to all concerned that Barker was entitled to extensions of time. After negotiations it was agreed that the work would be accelerated, and Barker would receive additional payment. HHJ Toulson QC held that the provisions of the sectional completion regarding liquidated damages were capable of continuing to have contractual force for the completion of each section, even though an acceleration agreement was in place, as it was common ground at the time of the agreement of acceleration that the liquidated damages clause would still have effect.

In *Ascon Contracting Limited* v. *Alfred McAlpine Construction Isle of Man Limited* [1999] Con LR 119, there had been delays due to a number of causes, and Ascon claimed for loss caused by acceleration measures it had undertaken. HHJ Hicks QC affirmed, *obiter dicta*, that 'acceleration' had no precise technical meaning. Acceleration, which was not required to meet a contractor's existing obligations, was likely to be the result of an instruction from the employer, for which he must pay. On the other hand, pressure from the employer to make good a delay caused by the contractor's own default was unlikely to be so construed. It was held that there could be both an extension to the full extent of the employer's culpable delay with damages on that basis, and also damages in the form of expenses incurred by the way of mitigation, unless it was alleged and established that the attempt at mitigation, although reasonable, was wholly ineffective.

In *Motherwell Bridge Construction Limited* v. *Micafil Vakuumtecchnik* (2002) TCC 81 Con LR 44, the issue of acceleration was addressed in a long and complicated judgment. One of the two claims presented by Motherwell was for the acceleration costs for work in relation to on-site fabrication and for hours worked in excess of normal working hours for a certain period of the project. HHJ Toulmin noted that, while Motherwell had not received any instructions to accelerate, it had generally been under pressure from Micafil to complete earlier and had employed additional resources to that end. There was no dispute that Micafil constantly urged Motherwell to increase its resources to meet the requested completion date. There was a term in the contract that, if unexpected delays and difficulties occurred, Motherwell was required to provide additional personnel at no extra cost at the request of Micafil in order to meet the required completion date. HHJ Toulmin held, *obiter dicta*, that the delays and difficulties came within the definition of 'unexpected', and Motherwell therefore could not succeed in recovering damages for this item.

## 4.12. Mitigation

The contractor has a general duty to mitigate the effect on its works of the employer's risk events. This duty to mitigate does not extend to requiring the contractor to add extra resources, or to work outside his planned working hours, in order to reduce the effect of an employer's risk event, unless the employer agrees to compensate the contractor for the costs of such mitigation. However, it can be argued that the obligation to progress the works diligently may require the contractor to take some positive action, and a failure to do so may result in liquidated damages being applied for the additional period of overrun which could have been avoided but for the breach.

An extension of time is a contractual remedy for acts of prevention and breach of contract by the employer and for events at the risk of the employer. It may therefore be thought that, if the remedy of extension of time is based on causation, the principles referred to as the duty to mitigate should apply. It is suggested that there are two situations to consider: first, when the contractor responds positively; and, second, when the contractor takes no positive action.

In the first situation, the contractor may react to the qualifying delay by making changes to his methods of working or sequence of working, or he may even accelerate the works. The issue then is whether he is entitled to recover the loss incurred by this reaction. The answer to this is that it depends on whether or not he was right to react as he did. It is suggested that, subject to the express terms of the contract, the contractor has no obligation or right to accelerate and is not entitled to recover additional costs incurred in acceleration measures to mitigate the effect of qualifying delays without an instruction from the client. As many contracts contain provisions for the granting of extensions of time and express terms for agreement of acceleration measures, the unilateral action by the contractor in giving priority to the fixed date for completion over the cost of working efficiently cannot bind the employer in those contracts. It is suggested that this interpretation can be expressed in terms of the reasonableness of the mitigation measures. It is normally not considered to be reasonable, particularly where there are sufficient contractual remedies available to the contractor, for the contractor to unilaterally decide to accelerate the works.

In the second situation, the contractor may not react to the qualifying delay, and the issue then is what minimum measures he is required to take in order to mitigate the effects of the qualifying delay, and if he fails to take those measures whether this affects the extent of his entitlement to an extension of time. It is suggested that, although the rules of mitigation do not generally apply to construction contracts with extension of time provisions and provision for recovery of time-related losses, the contractor will have some obligation to progress the works and this will involve an aspect of management of resources and planning of activities under the circumstances of the actual events. Although it is a matter of interpretation of the terms of the contract, it is suggested that such an obligation will usually be intended by the parties to apply equally to events causing qualifying delays.

## 4.13. The Delay and Disruption Protocol

Although not generally relevant to any particular contract terms, the Delay and Disruption Protocol is a document that is often referred to in the context of delay analysis. The Protocol was produced in the UK by the Society of Construction Law, and its aim was to create a practical environment where the number of disputes could be substantially reduced by the introduction of a transparent and unified approach to the understanding of programmed works, their records, and identifying the consequences of delay and disruption. The Protocol contained nothing new, but it explained in simple terms the enormous value of dealing contemporaneously with the effects of delay to progress via enforceable contract requirements for competent application of the critical path method and progress records that would enable the employer to manage his own risks.

Apart from providing recommendations and guidance to those involved in drafting contracts, the Protocol is intended to act as an aid to the interpretation of the delay and disruption provisions contained in standard form civil engineering and building contracts, and to act as a guide as to the manner in which contractors ought properly to prepare delay and disruption claims, and how judges and other tribunals ought properly to determine them.

The Protocol's position on core principles and other matters relating to delay and compensation can be summarised as

- preparing and maintaining programmes and records
- providing a mechanism for extensions of time during the course of the project
- dealing with disputed extension of time issues after completion of the project.

The Protocol recommends that a proper programme should be submitted by the contractor and approved by the contract administrator. The programme should show the manner and sequence in which the contractor plans to carry out the works. While recognising that the form of the programme will depend on the 'type and complexity of the project', the Protocol recommends that it should be prepared in the form of a critical path method using commercially available planning software. The Protocol further recommends that, once accepted, the programme should be updated electronically at intervals of no longer than a month. Each update should then be saved, the purpose being 'to provide good contemporaneous evidence of what happened on the project'.

The Protocol envisages that the updated programme will be the main tool for determining the amount of extension of time due. It further envisages that the programme will be brought fully up to date prior to the occurrence of an excusable delay event, and thus enable an accurate assessment of the amount of further time required when such an event occurs.

The Protocol describes float as 'the amount of time by which an activity or group of activities may be shifted in time without causing delay to a contract completion date'. The Protocol strongly recommends that ownership of the float, which may ultimately determine entitlement to an extension of time as a consequence of employer delay, should be adequately addressed in the contract. A contractor should not be automatically entitled to an extension of time merely because an employer delay to progress takes away the contractor's float for a particular activity. The Protocol's position is that an excusable delay should only result in an extension of time if it is predicted to reduce the total float on the activity paths affected by the delay. The Protocol distinguishes float as it relates to time and float as it relates to compensation. Whereas contractors traditionally take the view that float belongs to them, to be used as they see fit, the Protocol stipulates that float is a project resource, to be used when the project needs it.

The Protocol sets out various approaches to delay analysis, and these are, in order of preference.

- *Time impact analysis and windows analysis.* Time impact analysis can be considered under two headings: window analysis (also known as time slice analysis) and snapshot analysis. Window analysis is based on the analysis of the effects of delay events over the entire length of a project by looking at the events that have affected progress within 'windows' of the contract period sequentially. The duration of each window is not predetermined, but is frequently taken as being 1 month. At the end of each window, the as-planned programme is updated to take account of any delaying inefficiency which is the contractor's risk, any necessary logic or duration revisions because of mitigation measures undertaken, together with all excusable and/or compensable events during the period since the last update. The closing of a window in this way forms an as-built programme at the end of that window, and this effectively becomes the as-planned programme for the next window in the sequence.

- *Collapsed as-built analysis.* This method involves removing from the as-built programme identified excusable delays in order to show what the completion date would have been if those delay events had not occurred.
- *Impacted as-planned analysis.* This method adds an identified excusable delay event (or events), either as a separate activity (or activities), or as an addition to the duration of an existing activity (or activities), into the as-planned programme. The duration of the activity is derived (where possible) from the resource allowances on the as-planned programme. The impact of this added activity to the projected completion date is then compared with the original completion date to assess the extension of time due.

## Cases

*Amalgamated Building Contractors Ltd* v. *Waltham Holy Cross UDC* (1952)

*Ascon Contracting Limited* v. *Alfred McAlpine Construction Isle of Man Limited* [1999] Con LR 119

*Balfour Beatty Construction Limited* v. *The Mayor and Burgess of the London Borough of Lambeth* (2002) 1 BLR 288

*Balfour Building Ltd* v. *Chestermount Properties Ltd* (1993) 62 BLR 1 QBD

*British Steel Corporation* v. *Cleveland Bridge & Eng Co.* [1981] 24 BLR 100

*City Inn* v. *Shepherd Construction* (2007) CSOH 190

*City Inn* v. *Shepherd Construction* (2010) CSIH 68

*Fortec Constructors* v. *United States* 8 Cl. Ct. 490 (1985)

*Gaymark Investments Pty Limited* v. *Walter Construction Group Limited* [1999] NTSC 143

*Glenlion Construction Ltd* v. *Guinness Trust* (1987) 39 BLR 89 QBD

*H. Fairweather and Co. Ltd* v. *London Borough of Wandsworth* (1987) 39 BLR 106

*Henry Boot Construction (UK) Ltd* v. *Malmaison Hotel (Manchester) Ltd* [1999] 70 Con LR 32

*Henry Boot Construction (UK) Ltd* v. *Malmaison Hotel (Manchester) Ltd* (2000) CILL 1572 TCC

*John Barker Construction Ltd* v. *London Portman Hotel Ltd* (1996) 83 BLR 31

*John Doyle Construction Limited* v. *Laing Management (Scotland) Limited* (2002) CILL 1870

*Motherwell Bridge Construction Limited* v. *Micafil Vakuumtecchnik* (2002) TCC 81 Con LR 44

*Multiplex Constructions (UK) Ltd* v. *Honeywell Control Systems Ltd*, TCC 6 March 2007

*Natkin & Co.* v. *George A. Fuller Co.* 347 F. Supp. 17 (W.D. Mo. 1972)

*Neodox Limited* v. *The Borough of Swinton and Pendlebury* [1958] 5 BLR 38

*Peak Construction (Liverpool) Ltd* v. *McKinney Foundations Ltd* (1970) 1 BLR 111

*The Royal Brompton Hospital NHS Trust* v. *Hammond & Others* [2001] 76 Con LR 148, QBD (TCC)
*Wells* v. *Army & Navy Cooperative Society* (1902) 86 LT 764

**REFERENCE**

Society of Construction Law (2002) *Delay and Disruption Protocol 2002/2003*. SCL, Wantage.

**Delay and Disruption Claims in Construction**
ISBN: 978-0-7277-5705-0


doi: 10.1680/ddcc.57050.049

# Chapter 5
# Loss and expense

## 5.1. Delay-cost and disruption-cost claims

Participants in the construction and civil engineering industries often make the mistake of believing that time and money are directly linked, in that they assume that financial claim recovery can *only* be secured if an extension of time has been granted and, conversely, that it will *certainly* be due if an extension has been given. As noted in Chapter 4, in reality neither of these propositions is actually correct. Additional monies may be recovered even if prolongation is not present, and, conversely, a contractor may receive a comprehensive extension of time and still not be entitled to additional financial recovery. Therefore, it is advisable that notices for delay and notices for loss and expense are kept separate.

Delay-cost claims are claims for additional time-related costs associated with delays caused. Therefore, a delay-cost claim is a claim for delay costs where a contractor must show that the cause of the delay is one that entitles the contractor, under the contract, to payment for the extra costs incurred. To establish an entitlement to delay costs, a contractor must demonstrate that a delay to completion of the contract has occurred and must show that the cause of the delay is one that provides the contractor with an entitlement to extra payment, either under a term of the contract or for breach of contract by the employer. The claim must be supported by evidence of the facts on which it is based. To warrant the payment of delay costs, a delay must affect the critical path and delay completion of the whole of the works or a milestone.

Disruption-cost claims are claims for additional costs associated with disruption caused to the progress of or the execution of the works that do not affect the completion date of the project as a whole. In respect of a disruption-cost claim, a contractor must show that the cause of the disruption is one that entitles the contractor under the contract to payment for the extra costs incurred. To establish an entitlement to disruption costs, a contractor must demonstrate that a disruption to the progress or the execution of the works has occurred and must show that the cause of the disruption is one that provides the contractor with an entitlement to extra payment, either under a term of the contract or for breach of contract by the employer. The claim must be supported by evidence of the facts on which it is based.

## 5.2. Common loss and expense heads

The common heads of a loss and expense claim are

- on-site establishment costs (site overheads)
- head office overheads
- loss of profit
- increased costs
- working under different weather conditions
- cost of claim
- interest and finance charges
- disruption.

### 5.2.1 On-site establishment costs (site overheads)

Where a contract specifies a rate for delay costs that are applicable in the event that a contractor suffers delay, calculating delay costs is a straightforward exercise once the appropriate extension of time has been determined. The contractor is entitled to be reimbursed at the specified delay-costs rate for the period by which the contract time was extended, subject to any exclusions stated in the contract.

However, where there is no specified rate for delay costs, which is more usually the case, calculating a contractor's entitlement can be complex and time consuming. A contractor's on-site establishment costs (site overheads) are also commonly referred to as 'preliminary items'.

Additional expense under this heading is usually ascertainable from the contractor's cost records. Such costs may include

- site-based project manager
- site agent
- site foremen
- site quantity surveyors
- site engineers
- site offices
- site administration costs
- site stores
- welfare facilities
- craneage/forklifts
- general and small plant
- scaffolding/towers
- telephone
- electricity
- insurance.

Under this head of claim, what the contractor is seeking to establish is that, because of events that have occurred and for which the contract provides an entitlement to an extension of time, actual resources of the type listed above (amongst others), additional to those which otherwise would have been necessary, have reasonably had to be engaged on the project for a prolonged period.

Sometimes it is necessary for the on-site establishment costs to be broken down into delay costs (due to prolongation) and 'thickening' (or extra resource) costs (due to disruption).

The claim is usually for time-related costs, i.e. the additional cost associated with the prolonged use of a particular resource. For example, a 5-week delay will clearly affect the contractor's time-related costs, such as site supervision, cabins and hired equipment.

However, this may not always be the case. It may be that that the contractor incurs *additional* supervision costs without the overall completion date for the contract works being affected. For example, additional supervision costs may be incurred by the contractor during the original contract period, or a contractor may be delayed or disrupted by an instruction ordering additional work to non-critical items (e.g. scaffold to an area of work that was not on the critical path). In such circumstances, the contractor may be entitled to loss and expense incurred (i.e. thickening, as noted above), regardless of whether the additional work delayed the contractor in the completion of the works.

The prolongation part of the claim must relate to those periods when delay occurred, and this will not necessarily be the overrun period. For example, if a contractor had a contract period of 60 weeks, but was delayed in week 20 for 1 week, the site overhead costs would need to be claimed for week 20 (i.e. the week when the delay occurred) and not for week 61 (i.e. the week when the delay effect became apparent).

In most cases, the actual costs incurred by the contractor are to be claimed. A *pro rata* amount of the tendered allowance for on-site establishment costs clearly does not represent actual cost, and may therefore not be successful. In assessing such an on-site establishment cost claim, several points should be kept in mind.

- In the first instance it must be established that additional resources were required.
- The contractor needs to establish with sufficient particularity:
  - what was included in his tender, and
  - that the tender allowance was sufficient.
- Just because resources are engaged during a period of overrun does not mean they are additional. For example, if a contractor originally expected to have two foremen on site for 10 weeks but actually only had one foreman on site for 20 weeks, there would not (on the face of it) be any additional costs.
- The contractor needs to establish that the additional resources were required due to the reasons that he is relying on. This may be done in a couple of ways:
  - If the claim is for prolongation, then it is necessary to show resources were retained on site during the periods of delay for reasons outside the contractor's control. This is usually one of the easier points of the claim. However, the contractor does not need to have received an extension of time before he can pursue such prolongation costs during a period in excess of the original period. Conversely, not all extensions of time circumstances give rise to financial entitlement, and, even if they do, not all the additional resources may be recoverable.
  - A claim for on-site establishment costs may be pursued in the absence of an extension of time. If an appropriate resource is required for longer during the

original contract period, or more of a resource is required for the same period, then the costs of these should be recovered if they can be associated with a specific event. This may apply, for example, in the situation where a variation is issued that results in certain work activities which were programmed to be executed consecutively now needing to be carried out concurrently, thus requiring two or more work faces to be operating and two (or more) sets of supervisors, etc., where previously one would have been adequate.

- The contractor then has to establish the cost of the resources. With supervisory staff this would be ascertained using salary levels, pensions, national insurance, car costs, petrol, bonus, etc. If agency staff have been used then their actual costs will be recoverable, providing it was reasonable to use them.

On projects where the level of construction activity varies significantly between various stages of construction, the appropriate costs will be those that relate to the periods during which delay has occurred. To adopt average daily or weekly site overhead rates over the whole contract duration is not necessarily appropriate. The weekly cost of site overheads is related to the total work activity being undertaken on a construction site at any time, and not only to work on the critical path. Although a delay to critical work may occur at a period of maximum site activity, the effect may be to prolong only relatively few activities for an additional time. The appropriate rate for calculating site overheads is that related to the sequence of activities which were delayed, together with any consequential effects attributable to the delay.

Some non-critical activities may be delayed by a delay to critical activities on which they were dependent. These activities may still be non-critical, but they will be undertaken at a later time. This shift in time may not result in additional site overheads overall. It may merely cause the site overhead costs to be incurred at a later time. Site overheads are broadly related to the direct cost of the work to be undertaken. For projects with a high labour content, the cost of supervision will be greater than for projects with a high plant or material content. Less supervision is required for work where subcontractors are used compared with work requiring unskilled labour or using the contractor's own employees.

One method of assessing on-site delay costs is to evaluate the actual costs incurred. If cost information is provided by a contractor to justify a claim, then this must be audited to eliminate all costs that should be included in the direct costs of construction activities. Generally, there will be very few material costs within site overheads. Costs that are not time related would be deducted. For example, mobilisation and removal costs will be incurred irrespective of the contract duration. The exception could be where equipment or staff may be demobilised temporarily at the beginning of a long delay and remobilised at the end of the delay period in order to save the cost of retaining these resources on site for the duration of the delay period.

The actual cost of staff and labour will generally be provided by wages or salary sheets, and the cost of all external plant and services will be substantiated by invoices. These invoices should be inspected to ensure that no operating charges such as repairs or replacement parts, fuel or other incidentals are included. The delay cost is the equipment rental charge only. In some cases, there will be reduced hire rates or even no charge for standby or non-operational periods. Where contractor-owned plant is involved, invoices may not be available, and an analysis of the costs claimed will have to be made separately.

The importance of site records to assist in recognising excessive costs cannot be overstated. Where the client has maintained good site records, the checking of the actual times claimed against contractor records is the ideal way of establishing costs and the best way of accurately assessing the costs. A contractor is required to mitigate costs in the event of delay. Excessive expenditure due to poor management or inefficiency, if proven, is not recoverable. Specific instances of obvious waste or mismanagement would need to be identified in the costs or noted from site observations. Examples of actions that could be taken by contractors to mitigate losses include

- terminating the hire of plant not being used or hiring out plant
- laying off workers who are not productive, when this is possible
- increasing resources (e.g. additional workers or subcontractors), with a net benefit of a reduced delay impact
- working overtime
- reorganising work programmes and order of works.

To quantify a delay claim, a contractor must demonstrate that the delay caused damages, in the case of a breach of contract, or extra cost, in the case of a specific contract provision. The onus is on the contractor to prove that the costs claimed have been incurred, and that every effort has been taken by him to minimise these costs. If a contract does not include a prescribed delay-cost rate, then it is necessary to assess what delay costs are legitimate and to evaluate those costs.

### 5.2.2 Head office overheads

Head office overheads are sometimes called ‘off-site overheads’ or ‘establishment charges’. They are those administrative and management costs of running the head office of a contractor’s business over and above the site costs. Generally, head office overheads will include

- purchase and/or rent of office and yards
- maintenance and running costs
- transport and mobilisation costs
- company cars
- depreciation
- directors’ emoluments and expenses
- salary and other costs of staff
- administration costs
- legal and audit fees, etc.

Most contractors’ head office resources exist to support operations undertaken on site, and such head office overheads are normally taken as being recovered out of the income from the contractor’s business as a whole.

Where completion of one project has been delayed, a contractor may claim to have suffered a loss arising from the diminution of his income from the project, reducing the turnover of his business. Despite this, the contractor continues to incur expenditure on head office overheads, which he cannot materially reduce or, in respect of the project, can only reduce, if at all, to a limited extent. Therefore, but for the delay and disruption, the contractor's workforce would have had the opportunity of being employed on another project, which would have had the effect of contributing to the head office overheads and profit during the overrun period.

There is some authority that a claim on this basis is sustainable. In *J F Finnegan* v. *Sheffield City Council* (1988) 43 BLR 124, Judge Sir William Stabb said:

> It is generally accepted that, on principle, a contractor who is delayed in completing a contract due to the default of his employer, may properly have a claim for head office or off-site overheads during the period of delay, on the basis that the workforce, but for the delay, might have had the opportunity of being employed on another contract which would have had the effect of funding the overheads during the overrun period.

It should be noted that the entitlement to such general overheads is in itself an arguable point. However, the consensus of opinion is in favour, providing of course it can be established that such overheads would have been recovered were it not for the delay. That said, substantial claims of this kind are rarely made because most contractors are able to cope with delay on a particular contract using their existing resources, the cost of which is reasonably constant.

In terms of first establishing liability, the contractor must seek to prove that head office overhead costs have been increased as a result of a delay and disruption suffered, for which he is entitled to recompense under the contract. Examples may range from the cost of extra staff recruited because the particular project was in difficulties, to the cost of extra telephone calls and postage during the period of delay.

It is suggested that, in order to succeed, a contractor must provide evidence to show that

- the profit or overhead contribution was capable of being earned elsewhere at the time of delay
- the profit and overhead percentage is a reasonable one
- work of the same level of profitability and/or overhead recovery was available during the period of delay.

In terms of evidencing that there was other work available which, but for the delay, he would have secured, the contractor might seek to evidence this by producing declined invitations to tender, with evidence that the reason for declining was that the delay in question left him insufficient capacity to undertake other work.

The contractor might alternatively show from his accounts a drop in turnover and establish that this resulted from the particular delay rather than from extraneous causes. If loss of turnover resulting from delay is not established, the effect of the delay is only that receipt of the money is delayed. It is not lost.

If liability is established, then the contractor must establish quantum. When pursuing a claim for head office overheads, a subcontractor will frequently rely on the use of a formula to evidence such quantum. There are three formulae in common usage for the calculation of head office overheads (and, in addition, there are many other derivatives from these formulae). These formulae are

- the Hudson formula
- the Emden formula
- the Eichleay formula.

While the use of formulae has been heavily criticised by many commentators, there is judicial authority for their use in certain instances. However, whichever formula may be used, it must be noted that their use merely provides the means for assessing quantum. The fact must first be established that there was an actual loss, and that this loss flowed directly from the relevant matter relied upon.

In addition, when submitting claims for head office overheads, care needs to be taken to avoid any double recovery or overlap with other claims or payments obtained by the contractor, such as variations, which have been computed using the contract prices as a basis. Normally, in such a situation the valuation of the variation will include for recovery of an element of additional head office overheads.

### 5.2.3 Loss of profit

Although there has been some debate about whether loss of profit is recoverable as part of a loss and expense claim, if the words 'direct loss and expense' are to be interpreted as equivalent to the measure of damages at common law for breach of contract (which is generally accepted to be the case), then there can really be no doubt that a contractor can claim for the loss of profit that he would have earned on other contracts had there been no delay and disruption to the project in question, provided that such loss of profit was foreseeable.

Clearly, for a claim of loss of profit to be made, a contractor would need to show, as with head office overheads, that at the time of the delay he could have used the lost turnover profitably. The breach by the employer must be shown to have prevented the contractor from earning income which the contractor would have earned if the breach had not occurred. A claim for loss of profit does not, it is submitted, fail merely because the contract in question was unprofitable. The question is what the contractor would have done with the money if he had received it at the proper time. Even if, at that time, the contractor's business was making a loss, a sum analogous to loss of profit is, it is submitted, recoverable if the loss of turnover increased the loss of business.

Provided that the contractor has mitigated the loss, there may be a permanent loss in income which is recoverable from the client as damages. It is irrelevant whether the income lost is profit or overheads or categorised in any other way. It only has to be shown to be income lost. In such instances, the real cause of the loss of income is that the contractor is occupied because of the delays. The delay prevents the contractor from earning income. If, however, the employer pays for the contractor's resources during the delay period, the lost income is in effect reimbursed through those payments, and there is therefore a limited basis for a claim. A contractor's income will vary from time to time and, although the delay on one contract may reduce the income from that contract in one financial year, the income from that contract will be received in the next financial year, and there is often no permanent loss of income.

### 5.2.4 Increased costs

This head of claim usually arises on a fixed price contract (although it can equally apply where a fluctuation recovery has been agreed) where expenditure on resources due to increased costs during an extended period is sought.

It is submitted that the correct measure is the difference between what the contractor would have spent on resources and what he actually spent due to delays suffered. There are, however, a number of different methods of calculating this head of claim.

The best way is for the claim to be based on substantiated details of the level of costs that would have been expended on resources and the actual costs paid for those same resources backed up by invoices, etc. However, it is rare in practical terms that such an analysis could be economically undertaken. Therefore an alternative method would be to use some type of formula approach.

Using a formula approach, the calculation would broadly be as follows.

- Ascertain the value of resources in the original tender but using actual work values.
- Find out the level of inflation for such resources over the original contract period, normally by using some standard published indices.
- Calculate how much the cost of resources would have increased during the original contract period based on the indices obtained. This will then be taken as the amount deemed to be allowed in the rates.
- Undertake the same calculation but using the indexed inflation figure over the extended contract period.
- The extra over-cost between the above two calculations will constitute the increased cost claim.

### 5.2.5 Working under different climatic conditions

This heading amounts to a different aspect of loss of productivity and arises from the need to work in less favourable climatic conditions. This is a recoverable head of claim (*Bush* v. *Whitehaven* (1888) 52 JP 392).

What the contractor has to establish in this case is that, for reasons providing an express contractual entitlement, he has had to undertake work in climatic conditions less favourable than would otherwise have been the case. It is not enough to show that the work was undertaken during a period of less favourable climatic conditions, it must also be shown that this fact did actually affect the production level of the work in question.

### 5.2.6 Cost of claim

Where the preparation of a claim is undertaken by a contractor's in-house staff, this cost will often be partly (or completely) reflected in the contractor's loss and expense claim for site overheads or head office overheads. However, often a contractor may employ a consultant to assist in the preparation of a loss and expense claim, and the general rule is that the costs incurred in taking this action are generally not recoverable, except and until the matter proceeds to arbitration (*James Longley* v. *South West Thames* (1984) 25 BLR 56) or litigation.

The reasoning behind disallowing this head of claim is that the contractor is obliged, under the contract, to provide information, etc. to resolve financial issues, and therefore the submission of a claim and the work involved must have been included in his price. It may be the case, however, that such monies might be recovered if the employer requested further information from the contractor, which are judged to be beyond the contract requirements.

### 5.2.7 Interest and finance charges

When a contractor incurs loss and expense, this has to be financed by him either from his own capital resources or by increased borrowing. In either case, it is clear that the use of money costs money. Depending on the level of inflation (and the interest rates applicable at any point in time), and the amount outstanding in respect of the loss and expense claim, this matter will either be a relatively small matter or will have major implications. In either event, it is a matter of great importance to the contractor how such interest or finance charges should be recovered.

Interest in respect of monies owed is often dealt with as an express term of the contract. Therefore, if it is proven that monies were outstanding in respect of the non-payment or under payment of a loss and expense claim, simple interest would be applied in line with the contract in question. In the alternative case where monies were borrowed, finance charges, or the cost of being stood out of one's money, is a recoverable head of a loss and expense claim in any event. This was established in the Court of Appeal case of *F G Minter* v. *WHTSO* (1980) 13 BLR 1, CA.

In respect of common law damages, to recover finance charges it is necessary for the claimant to show that the finance charges fell within the second limb of *Hadley* v. *Baxendale* (1854) 9 Exch 341 (for detail, see Chapter 3), i.e. that the finance charges are special damages that were in the contemplation of the parties at the time when the contract was formed (see: *President of India* v. *Lips Maritime* [1988] AC 395, HL; *Holbeach Plant Hire* v. *Anglian Water Authority* (1988) 14 Con LR 101).

### 5.2.8 Disruption

This claim head relates to the inefficient or increased use of labour and plant resources and is commonly referred to as a 'disruption claim'. Delay and disruption can lead to increased expenditure on plant and labour in two ways: it may be necessary to employ extra resources; and existing resources may be used inefficiently.

Disruption is often treated by the construction industry as if it were the same thing as delay. The two issues are, however, entirely separate matters. Delay is lateness, whereas disruption is loss of productivity, disturbance, hindrance or interruption to the contractor's normal working methods, resulting in lower efficiency. Disrupted work is work carried out less efficiently than it would have been if not for the cause of the disruption. Disruption compensation is only recoverable to the extent that the employer caused the disruption. Most standard forms of contract do not deal expressly with disruption, but disruption may be claimed as a breach of the term generally implied in construction contracts that the employer will not prevent or hinder the contractor in the execution of his work.

Disruption costs may be distinguished from delay costs by virtue of the fact that the latter are a function of time and the former are essentially productivity related. In a disruption claim, contractors claim that they could not achieve their planned output because of employer actions or other causes not their responsibility, and hence that the damages or extra costs are payable.

Causes of disruption can be broken down into either external and internal causes. External causes of disruption are generally not related to the project itself, and will often fall into the *force majeure* category. These causes will also include government acts, such as the passing of new regulations, changes to taxes and new laws. Such events will generally involve certain rights of compensation to be passed on to the contractor. External causes can also result from acts of God, such as earthquakes, etc.

Internal causes of disruption can be causally attributed to the project itself, its planning and design and the manner in which the works are performed. Internal causes of disruption can be further broken down into: technical causes, which include changes in design, design errors and construction errors; and economic causes (e.g. difficulties in accessing requisite materials, labour or skills) and financial causes (e.g. lack of employer funds, material or labour cost increases and interest rate rises).

It will usually be expected on a complex project that a certain amount of uncertainty and rework will be expected at various stages of construction. Even when the project is going well, normal disruptions, made by both the contractor and the employer, will involve a certain amount of rescheduling and planning, and additional costs to rectify. Despite the fact that these costs are built into the initial tender and will be absorbed without affecting the time frame or budget, it is possible to drastically underestimate the costs of such factors.

An example of disruption is where the employer ordered a contractor to cease work on a particular activity for frequent short periods (e.g. to provide for some necessary operating function of a plant or building), and the need for such stoppages was not specified in the tender documents. Another example would be where the employer ordered urgent variations, and groups of workers had to continually move from one activity to another at short notice, being unable to develop optimum productivity on a particular work activity. Hence, a disruption claim must identify specific events that are breaches of contract by the employer or events for which the contract specifically provides for extra costs.

To justify a disruption claim, a contractor must establish that actual progress of the work has been interrupted and that the cause of the disruption was either a breach of the contract by the employer or an action for which the contract provides for the reimbursement of extra cost. Even if the employer does cause disruption, this may not result in an entitlement to additional payment. It may be that the contractor failed to comply with certain contractual requirements and is therefore not entitled to reimbursement of the disruption costs. A standard normal requirement for the validity of a disruption claim is that the contractor must give notice of information required at the time of knowing of the disruption. The contract usually will provide the mechanism for the notification with regard to the type of information required and to whom the notice should be addressed.

The practices, which are determinant to success or failure of entitlement to contractor compensation due to disruption, can be summarised as below.

- The work that has been affected must be clearly identified, and the work activities that were affected by the disruption must be specified. The extra expense incurred must be explained.
- The contractor must show that the event leading to the disruption and financial loss was either a breach of contract, or an event provided for in the contract for which the employer is to be made financially liable to the contractor.
- It must be shown that actual work progress has been negatively impacted. It is not sufficient to show that planned future work has been impacted, as such uncertainties may never materialise.
- The contractor must quantify the disruption costs using a selected method of quantification. The principle that applies is that the extra costs incurred, compared with the costs that would have occurred had the disruption not occurred, are recoverable by the contractor.
- The contractor sets out what the actual costs would have been had the disruption not occurred. This provides the comparator to be used in the calculation.
- The contractor must show that he has taken all reasonable steps to mitigate his loss, such as returning leased equipment, working on other parts of the project that were not affected by the disruption and redeploying expensive resources so that they were not unnecessarily sitting idle.

This head of claim in particular may be difficult to ascertain due to a lack of records and the difficulty of establishing the nexus of cause and effect. However, just because it is difficult to establish, it does not mean that a contractor should not submit a claim. Equally, it does not mean that the contractor should not make a reasonable assessment (see the comments on global claims in Chapter 3).

As noted in Chapter 3, the global claim approach (particularly in the UK) should be used very much as the exception rather than the rule. It is applicable and permitted only where the events are so numerous, complex and/or interrelated that no amount of analysis or record keeping would establish or evidence the connection of cause and effect on individual matters. However, it is not good enough that such analysis cannot be done simply because the contractor has not kept reasonable records.

There is a number of methods of assessing this head of claim, and few people agree on the approach to use. However, in all cases, good records are essential. Ideally, what should be established is that, *before* a particular event occurred, productive resources were achieving a certain level of productivity or financial income for the contractor. However, *after* that particular event occurred, and indeed because of it, the same level of productivity or financial income could not be achieved.

Take, for example, the case of a bricklaying contractor. In the contract, the contractor is told that he will be required to build wall A, followed by wall B, followed by wall C. In his tender, the contractor allowed for an output of 60 bricks per hour per bricklayer.

- Week 1: things progress well. The bricklayers worked on wall A only, as planned; the output achieved was 65 bricks per hour. The contractor was exceeding his tender expectations, and this illustrated the sufficiency of his tender.
- Week 2: things start to go wrong. The sequence of work changed, and the contractor was instructed by the employer to switch his bricklayers from wall A to wall B to wall C, often all in the same day. As a consequence, the output of the bricklayers at the end of the second week reduced to 50 bricks per hour.

In such a case, the disruption claim would be based on the loss of production caused by the disruption event. In other words, the contractor expected to achieve 60 bricks per hour (or the contractor might argue that he actually achieved 65 bricks per hour) before the disruption event occurred; but because of the disruption event the contractor's bricklayers only achieved 50 bricks per hour. Although this is a very simple example, the principle remains the same for much more complicated claims.

The above approach in large part defeats the point that is often raised in respect of disruption claims that there was an insufficiency within the contractor's tender. This argument is often used when the contractor compares his actual output with the output allowance that he made in his tender. Where a contract does not prescribe a method for evaluating disruption costs, a contractor might make an ambit claim for the difference between the amount allegedly allowed in the tender and the actual cost of the work performed. Such an approach falsely assumes that the amount allowed for the work at the time of tender was totally correct and there were no inefficiencies in the contractor's management of the construction operations.

A contractor must quantify the disruption costs once he has established that loss of productivity has occurred and caused a delay. This involves comparing the actual cost with what the cost would have been had it not been for the disruption. The contractor must demonstrate that the latter cost is reasonable, although it is hypothetical to some extent. In making such claims, a contractor must also establish that everything reasonable has been done to minimise the cost of the disruption (e.g. that hired machines were not left idle on site when the hire could have been terminated). Cost details of the affected resource may then be compiled from the site accounting records. On this basis, any resulting disruption claim will be in respect to actual loss and expense incurred, instead of by reference to tender allowances.

As noted above, there are several types of approach that may be applicable in respect of a disruption claim, depending on the nature of the work, the circumstances and the records available. The main approaches used are

- to show that resources were standing idle when they should otherwise have been working, thus taking longer to do the same work
- to illustrate how the work of a particular trade has been prolonged and to apply a disruption factor in the same proportions to the cost of the trade
- to estimate a disruption factor
- to do an overall global claim based on costs of all trades assessed in the tender, and to claim the difference between that and the actual costs
- a mini global claim using the same approach as above, but targeting specific trades
- to illustrate the resources employed in producing specific and evidenced amount of work prior to an event, and to claim the extra resources for doing equivalent work after the event
- to show that more expensive resources were required to do the same value of work because of an event, even though the level of resources was the same (i.e. agency resources, emergency hire of plant, expensive gangs, etc.).

There are various methods to quantify loss of productivity in a delay and disruption claim. These methods, each having its own merits and limits, include the four described below.

#### 5.2.8.1 The industry standard approach

The industry standard method of quantifying labour productivity is not always favoured by courts due to its hypothetical nature. Trade groups for various contractors have published productivity tables that show how various job-site conditions can affect labour productivity. Particular conditions are scored in terms of their effect on productivity. For example, a contractor can use certain published factors and ascertain how much his productivity was affected by the disruption caused by the employer. These measurements are then quantified to show the loss of efficiency and the disruption claim amount. Employers contend that industry standards are factually distinguishable from the conditions actually experienced on their particular sites. Nonetheless, industry standards have been used successfully as a way to measure productivity when actual data are missing or otherwise unavailable.

#### 5.2.8.2 The total cost and the modified total cost methods

The total cost method is based on the premise that the resulting project is a cardinal change from what was originally contracted, i.e. the current project is fundamentally different from the project envisioned by the contract, or when other methods of computing losses cannot be done. Once a cardinal change, or abandonment, has been established, the contractor is freed from the terms of the contract and is allowed to recover the reasonable value of labour and materials, plus reasonable mark-ups for overhead and profit, less what was previously paid.

The modified total cost method refines the total cost method by subtracting from the total costs any costs incurred by the contractor due to his own inefficiencies.

The main criticism of the total cost and modified total cost methods is that they can be used by contractors to hide losses not caused by the employer, such as losses due to the contractor's tender errors or defective project management. Employers frequently seek to bar the total cost and modified total cost methods on these grounds. Employers also contend that the contractor cost records are sufficiently detailed so that, during the course of a project, the contractor could have tracked his costs to show an actual causal relationship between the employer's actions and the contractor's loss of production. Employers argue that to the extent the contractor failed to adjust his accounting system to track job costs, the contractor should be barred from using these methods.

#### 5.2.8.3 The jury verdict method

This method allows the 'trier of fact' to determine recoverable delay and disruption damages. To apply this method, a party must present

- a clear proof of injury
- an indication that there is no more reliable method of computing damages
- evidence of sufficient weight of fact to make a fair and reasonable approximation of damages.

Frequently, courts use evidence submitted in support of other quantification methods, such as the methods described above, to derive a jury verdict calculation. Employers contend that the jury verdict method should not be used because the claimant contractor failed to meet its burden of demonstrating damages with reasonable certainty. Employers also contend that the jury verdict method, in essence, amounts to nothing more than allowing the jury or the court to make a 'guess' regarding what the contractor damages are, and this process frees the contractor from his normal and customary burden of showing breach, causation and damage.

#### 5.2.8.4 The measured mile analysis

The measured mile approach, which involves comparing the actual cost with what the cost would have been were it not for the disruption, is one of the main proven methods a contractor applies in order to quantify costs once it has been established that loss of productivity has occurred and caused delay. It involves an evaluation of disruption carried out on the basis of a comparison of productivity prior to the disruptive events taking place with the productivity achieved during the period of disruption. A comparison of outputs is usually made by assessment of sums certified within interim payment certificates. The productivity rate for a period of disruption is quantified in lost worker-hours, which are multiplied by an hourly rate to find the loss of labour productivity or the disruption claim amount.

In its purest application the measured mile measures two different periods of productivity for the same type of work performed under the same type of physical conditions on the same project. On this basis, any resulting disruption claim will be in respect of actual loss and expense incurred instead of by reference to tender allowances. Employers may disagree with a measured mile calculation by claiming that the baseline productivity measure (i.e. the measure of the period unaffected by disruption) is faulty. A baseline productivity measure can be inaccurate if the delayed period versus the planned period comparison is not an accurate comparison.

In the case of *Whittal Builders Company Ltd* v. *Chester-Le-Street District Council* (1985) 11 CLR 40, the court had to decide the most appropriate method of evaluating the disruption that had resulted from the problems relating to the handover of properties. Of the variety of methods presented before the court, the measured mile method of evaluating disruption was the favourite method approved. Mr Recorder Percival said:

> Several different approaches were presented and argued. Most of them highly complicated, but there was one simple one – that was to compare the value to the contractor of the work done per man in the period up to November 1974 with that from November 1974 to completion of the contract. The figures for this comparison, agreed by the experts for both sides, were £108 per man week while the breaches continued, £161 per man week after they ceased. It seemed to me that the most practical way of estimating the loss of productivity and the one most in accordance with common sense and having the best chance of producing a real answer was to take the total cost of labour and reduce it in the proportions which those actual production figures bear to one another – i.e. by taking one-third of the total as the value lost by the contractor. (This roughly being the difference between the productivity when work was not disrupted, i.e. £161 per man week and £108 per man week when work was disrupted). I asked both [counsel] if they considered that any of the other methods met those same test as well as that method or whether they could think of any other approach which met them better than that method, In each case the answer was 'no'. Indeed, I think that both agreed with me that that was the most realistic and accurate approach of all those discussed. But whether that be so or not, I hold that it is the best approach open to me, and find that the loss of productivity of labour and in respect of spot bonuses which the plaintiff suffered is to be quantified by adding the two together and taking one-third of the total.

In the case of *John Doyle Construction Ltd* v. *Laing Management (Scotland) Ltd* (2004) Scots CS 141, the court was asked to express a view as to whether a measured mile approach is, in principle, acceptable. A part of the claim related to disruption, and the evaluation was produced by comparing labour productivity actually achieved on site when work was largely free from disruption with labour productivity achieved when work was disrupted. It was decided that this method of evaluating a claim was not a total cost claim and, in principle, was acceptable.

## Cases

*Bush* v. *Whitehaven* (1888) 52 JP 392
*F G Minter* v. *WHTSO* (1980) 13 BLR 1, CA
*Hadley* v. *Baxendale* (1854) 9 Exch 341

*Holbeach Plant Hire* v. *Anglian Water Authority* (1988) 14 Con LR 101
*J F Finnegan* v. *Sheffield City Council* (1988) 43 BLR 124
*James Longley* v. *South West Thames* (1984) 25 BLR 56
*John Doyle Construction Ltd* v. *Laing Management (Scotland) Ltd* (2004) Scots CS 141
*President of India* v. *Lips Maritime* [1988] AC 395, HL
*Whittal Builders Company Ltd* v. *Chester-Le-Street District Council* (1985) 11 CLR 40

**Delay and Disruption Claims in Construction**
ISBN: 978-0-7277-5705-0


doi: 10.1680/ddcc.57050.065

# Chapter 6
# Case study: extension of time and related losses

This case study seeks to demonstrate an entitlement to an extension of time and related losses in a 'real-life' situation, and the format (although obviously not the content) of this case study can be used as the basis of any such claim.

It is useful to start any claim submission with an executive summary. The executive summary provides an overview of the entire claim and makes it simple for anyone reading the claim to understand the basis of the claim and what is being sought as redress from the claim.

On the basis that most claims are made pursuant to a contractual right, it is normal to follow the executive summary with details of the contract and the relevant contract terms that will be relied upon within the claim submission. A background to the claim submission then needs to be provided, followed by a narrative of events.

The key part of the submission is to provide a nexus of cause and effect, and this is normally produced in respect of extensions of time by way of a delay analysis, and in respect of prolongation costs by way of an evaluation of those costs.

In line with the above comments, the case study below is set out under the following headings

- executive summary
- the contract
- background
- narrative of events
- delay analysis
- evaluation of prolongation costs.

## 6.1. Executive summary

This is a claim by a contractor, Contractor Limited (the Contractor), for an extension of time and related losses for the amount of $57 779 070 which was incurred due to delays and disruptions to the progress of the works caused by the employer, Employer Limited (the Employer), and the Employer's directly employed infrastructure contractor (the Employer's Infrastructure Contractor).

The object of this claim submission is to detail the circumstances, which were totally beyond the control of the Contractor, whereby unavoidable delays and disruptions to the contract occurred and which subsequently caused prolongation in the execution of the works and financial losses to the Contractor. Entitlement to reimbursement of these costs is fully covered under the terms of the contract.

The Contractor was prevented from carrying out its duties to complete the works in the pre-set agreed and signed upon timeframes. Consequently, pursuant to the express wording of the contract conditions, the Contractor is entitled to receive a commensurate extension of time and related financial compensation for the resulting prolongation of the time for completion.

This claim submission establishes the Contractor's entitlement to an equitable extension of time and related losses. It demonstrates the inability of the Employer and the Employer's Infrastructure Contractor to hand over the site at the required times, and also demonstrates that the Employer has caused other events leading to additional delays and disruption to the progress of the works. The following evidence will be adduced.

- A chronological record of delay events caused by the Employer. (For the purpose of this case study, the letter reference, the date and a brief summary of each correspondence is described. It is advisable that when writing a claim, the claimant must elaborate more on this section in terms of the effect of each correspondence and its impact.)
- A Time Impact Analysis. (For the Time Impact Analysis given in this case study, a summary of the impacted finish date of each zone is presented. It is advisable that a more detailed study is required, with a full appendix showing the methodology and programming of the Time Impact Study.)
- Cost implications resulting from the identified delay events.

The total amount of this claim is the cumulative sums of the delay amount, the disruption amount and the acceleration amount, which equals $57 779 070.

## 6.2. The contract

On 26 April 2005, the Contractor entered into a contract with the Employer for the redevelopment of existing housing, business and service quarter facilities, and the execution of new housing family units, recreational and office buildings with supporting utility networks, and associated external works to a project known as Project A. The contract start date of Project A was 4 June 2005 and the contract completion date was 13 October 2008; the contract period was therefore 1227 days. The contract sum was $259 200 000.

The Contractor's right to receive compensation in terms of time and cost are governed by the wording of the particular clauses contained in the particular and general conditions of the contract agreement with the Employer. The clauses that are particularly relevant to the events in this statement of claim are reproduced in this section.

**Clause 7 – Employer's Obligations**

7.1 The Employer shall, in accordance with and subject to the terms and conditions of this contract.

7.1.1 Obtain all permits, licenses, and other governmental authorisations which official procedures require to be obtained in the Employer's name and which are necessary for the performance of the work.

7.1.2 Allow the Contractor partial and progressive access, subject to the Employer's normal security control and safety procedures, to the portions of the work site as required in the scope of work for the orderly performance of the work.

7.1.3 Obtain any rights-of-way that are determined by the Employer to be required for the performance of the work.

7.1.4 Perform all other obligations required of it by the terms of this contract in such time and manner as to facilitate the orderly execution of the work.

**Clause 9 – Work Commencement, Execution and Completion**

9.1 The critical milestones and/or time for completion of work will be extended if the Contractor is delayed or impeded in the performance of the work by causes expressly set out in this contract that permit an adjustment of the critical milestone dates and/or time for completion of work, or by a breach of contract or 'act of prevention' by the Employer, by such reasonable period as fairly reflects the delay or impediment sustained by the Contractor.

9.2 If the Contractor intends to apply for an extension of a critical milestone and/or time for completion, the Contractor shall give notice to the Employer of such intention as soon as possible and in any event within 28 days of the start of the event giving rise to such delay, together with any other notice required by the contract and relevant to such event. Within 28 days of the date of issue of such notice, the Contractor shall submit full supporting details of his application with an analysis of the delay impact on his work schedule as well as any particulars and explanations that may be requested by the Employer. The Contractor shall make his best efforts to mitigate delays that arise. The Employer will reject any application for an extension of time for completion that does not comply with the requirement of this sub-clause.

**Clause 19 – Claims Settlement: Disputes**

19.1 The Contractor shall inform the Employer, in writing, within thirty (30) days following its occurrence or discovery, of any item or event which the Contractor knows, or reasonably should know, may result in a request for an extension of time for completing the work or any separate portion thereof, or for additional or reduced compensation under this contract. The Employer and the Contractor shall endeavour to satisfactorily resolve the matter. Should it not be disposed of to the Contractor's satisfaction, the Contractor shall forthwith deliver a written notice of claim complete with all supporting documentation in triplicate to the Employer at its address.

19.2 Notice of any claim by the Contractor against the Employer for an extension of time for completion of work or any separable portion thereof, or for additional compensation of any kind under this contract, shall be set forth in writing by the

Contractor and filed within thirty (30) days after failure to agree pursuant to Sub-clause 19.1 hereof. This notice shall contain a written analysis of all elements of the claim accompanied by itemised supporting data identifying, to the extent practicable, the effect on time for completion of work or any separable portion thereof and the amount of additional compensation claimed of work or any separable portion thereof and the amount of additional compensation claimed by the Contractor. Failure by the Contractor to so file such written analysis with supporting data within said thirty (30) days period shall be deemed conclusively to be a waiver of all of the Contractor's rights to any such extension of time for completion or to additional compensation.

19.3 Provided also that where an event has a continuing effect such that is not practicable for the Contractor to submit detailed particulars within the period of thirty (30) days referred to in Sub-clause 19.2 above, he shall nonetheless be entitled to an extension of time or additional or reduced compensation provided that he has submitted to the Employer interim particulars at intervals of not more than thirty (30) days and final particulars within thirty (30) days of the end of the effects resulting from the event. On receipt of such interim particulars, the Employer shall, without undue delay, make an interim determination of extension of time or additional or reduced compensation and, on receipt of the final particulars; the Employer shall review all the circumstances and, as appropriate, determine an overall extension of time or additional or reduced compensation in regard to the event.

19.4 Within a reasonable time of receipt by the Employer of the Contractor's written analysis of the claim with supporting data, the Employer shall issue to the Contractor a written response giving the Employer's determination detailing the terms of settlement or rejection of the claim.

19.5 Should the Contractor and the Employer be unable to agree upon a settlement of any claim, the matter shall be treated as an unresolved dispute.

19.6 Notwithstanding any other provisions, if the Contractor fails to comply with any of the provisions of the contract, in respect of any event, act or omission of whatsoever nature which, in the opinion of the Contractor fairly entitles him to an extension of the time for completion and/or additional payment, then such failure shall constitute on the part of the Contractor a definitive and irrevocable waiver of any entitlement and the Contractor shall be effectively barred from raising any claims arising from such event, act or omission thereafter.

19.7 Should any dispute arise between the Employer and the Contractor during the Contractor's performance of work, the Contractor shall, unless the Employer directs otherwise, continue to perform work and any additional work which the Employer may direct the Contractor to perform.

## 6.3. Background

The Contractor was prevented from carrying out its duties and obligations to complete the works as a result of the Employer and other entities, such as the Employer's Infrastructure Contractor, failing to complete their preparatory and infrastructure activities in the agreed and signed upon timeframes.

Consequently, pursuant to the express wording of the contract conditions, the Contractor is entitled to receive a commensurate extension of time and related compensation for the resulting prolongation to the time for completion. The Contractor incurred significant increased costs due to delay and disruption for reasons wholly beyond its control and responsibility. This statement of claim is for prolongation costs incurred by reason of the delays caused by the Employer and the Employer's Infrastructure Contractor.

From the outset, the completion of Project A, in the agreed time period of 1227 days, was a very challenging undertaking, and the completion of the different milestones of Project A was critical.

The Contractor has a well-deserved reputation as being a contractor who can complete large-size projects within demanding time periods without compromising the quality of the end product. The Contractor was chosen by the Employer for this reason, as well, presumably, because of its competitive pricing level.

The Contractor prepared a tender for Project A and, as a result, it was thoroughly conversant with Project A as originally designed, planned and tendered. Although the Contractor's familiarity with similar size projects undoubtedly assisted him in minimising the scheduled impact of variations by the Employer or for other delays that the Contractor was not liable for, that familiarity could not and did not alert the Contractor to the likelihood or magnitude of such delays and disruptions as ultimately were initiated by the Employer and others under the direct control of the Employer.

Clearly, under these very strict and unyielding contractual parameters, it was reasonable for the Contractor to assume that the different zones of Project A would be made available to carry out the works as the contract stipulated and that the handing over process would be effected in a planned and organised fashion. However, because of the fact that other parties involved in Project A (beyond the control of the Contractor) did not carry out their duties properly, and did not complete their tasks on time, the Contractor's reasonable assumptions were not met by the actions of others.

The Employer's objective was to obtain an excellent quality of work while maintaining its original budget. However, the Employer found itself dealing with a number of issues such as delays in the completion of the works due to its own directly employed Infrastructure Contractor's late execution of its parts of the works; the relocation of certain facilities and items; the unforeseen conditions resulting from the fact that the demolition and removal of debris was not properly executed by other contractors directly employed by the Employer; the flooding of the water well and existing utilities; all of which resulted in disruptions and delays to an already very challenging construction period. It is important to note that the Contractor offered remedial measures for certain delay items, but the Employer prolonged the reply to these offers and subsequently additional delays were encountered.

The Contractor was continuously requested to mitigate the delays caused by the Employer and others outside of the Contractor's control, but was not paid any additional costs for doing so, and was not awarded any extensions of time.

Acceleration measures were undertaken by the Contractor in order to keep Project A progressing with the least disruption possible, and to permit completion of the works at a date earlier than would normally have been the case as a result of the impacted delays.

The Contractor partially overcame numerous difficulties, including the delays initiated by the Employer's directly employed Infrastructure Contractor, who did not perform his duties according to the agreed schedule, and the Contractor was continuously working with the aim of mitigating delays and completing Project A with minimum time impact.

It is well understood and accepted within the construction industry that proceeding with the works on a 'fast-track' basis necessarily and inevitably disrupts a contractor's schedule of work and performance, which results in increased costs owing to lost productivity and inefficiency. The Contractor fully and faithfully cooperated with the Employer in performing the works on a 'fast-track' basis, without waiving any rights to recover the additional costs incurred due to the departure from its planned method of performance.

As noted above, the contract start date of Project A was 4 June 2005 and the contract completion date was 13 October 2008.

The site preparation works were awarded to an Infrastructure Contractor who was directly employed by the Employer.

The contract stipulated a specific timeline, including starting date and completion, for each zone in Project A in order to allow the Contractor to perform the works in a planned and orderly manner. Because of the delays and disruptions as set out above, the most dramatic loss of production experienced by the Contractor was the disruption which affected the mobilisation of crews and relocation of assigned resources, as well as the efforts in dealing with such disruptive and unknown conditions on a daily basis to accommodate the site-availability needs instead of concentrating on the project needs.

For any project of this size to be successfully completed within an ambitious time frame, the site availability is one of the main factors which affects the project. The Employer should have ensured that the infrastructure works were completed before it entered into a contract with the Contractor, particularly as it was clearly aware that the completion of such preceding infrastructure works was critical to the progress and completion of the construction works to Project A by the Contractor.

Unfortunately, the said site preparatory works were not completed as required, and this prevented the Contractor's commencement of the works in several zones and subsequently caused delays and disruption. Frequently, this required the hold up of certain zones of work, and the repeated re-sequencing of the Contractor's resources between zones unaffected and zones being delayed. Common sense and ordinary experience in the construction industry is that worker productivity and, consequently, the cost of the works are dramatically affected by such unanticipated conditions. This was not the basis on which the project was put out to tender by the Employer or bid for by the Contractor. Therefore, the Contractor's significant increased costs are directly attributable to the project conditions under which it was forced to perform for a substantial portion of the contract along with the disruption to the reallocation of its resources on a fast-track construction project.

As in any project, the availability of the site is, under normal circumstances, one of the critical activities in any programme. The Contractor attempted to mitigate these delays by working concurrently with the Employer's directly employed Infrastructure Contractor in several zones upon receiving access to certain parts whilst waiting for access to other parts. Unfortunately, the Contractor was deprived of executing its initial plan and had to wait for others to complete their tasks, and such process resulted in extensive delays and disruptions to the follow-on site activities.

The disruption caused the following ripple effects.

- Uncontrolled actions by others resulting in unplanned and belated handing over of the site, and in many instances at intermittent stages.
- Delays in completion of a number of critical activities, thereby impacting the completion date of Project A.
- Out-of-sequence work resulting in loss of labour productivity and equipment production, and therefore causing further disruption.
- The Contractor's inability to plan ahead, order materials and/or allocate resources in a timely manner.

## 6.4. Narrative of events

In order to prove causation, the Contractor provides a summary of the delays and disruptions suffered for each affected zone. In addition, all relevant correspondence with the Employer is listed.

### 6.4.1 Zone 1BC

The Contractor was not allowed to start its work until it met security requirements for Zone 1BC (not within the Contractor's original scope of work). Moreover, the Employer proposed to relocate the public toilets, which constitute part of the works to Zone 1BC, to another location within the premises. These events led to delays in the progress of works and, ultimately, the delayed completion of related areas. The Employer's requirement was to erect a permanent fence, and that the Contractor would have to defer construction until it received the area, and the Contractor would then resubmit a proposal to the Employer to reflect the permanent fence requirement. It should be noted that the erection of the permanent fence prior to construction created construction problems relating to the foundation of parts of the structure of the Zone 1BC.

List of correspondence: GM/L-0202-06 on 29 November 2006; PMC-398-0576-06 on 25 December 2006; GM/L-0246-07 on 3 May 2007; GM/L-0438-07 on 23 August 2007; PMC-398-0575-08 on 1 September 2008; GM/L-0011-09 on 7 January 2009; PMC-398-0052-09 on 27 January 2009; PMC-398-0181-09 on 15 November 2009.

### 6.4.2 Zone 1E

The Employer's directly employed Infrastructure Contractor was late in executing the demolition works (not within the Contractor's scope of works) and delayed the handing over of some areas, which affected the progress on site, leading to subsequent delays and prolonging the time for completion. Moreover, these delays caused disruption, unorganised resource allocation and additional associated costs (including lower productivity, idle time, etc.).

List of correspondence: GM/L-0118-06 on 7 November 2006; GM/L-0139-07 on 3 December 2007; GM/L-0173-08 on 21 December 2008; GM/L-0057-9 on 21 March 2009.

### 6.4.3 Zone 2A1

The handing over of Zone 2A1 was not complete and demolition works were still ongoing, and this late handing over caused delays to related works such as the construction of the villas.

List of correspondence: GM/L-0531-07 on 24 October 2007; GM/L-0554-07 on 3 November 2007; PMC-398-0023-09 on 13 January 20099.

### 6.4.4 Zone 2A2

Zone 2A2 was not vacated by others, and this led to a delay in the handing over of that area. Subsequently, progress of works was delayed. On 4 December 2007, the Contractor requested the disconnection of utilities prior to the commencement of demolition works. A portion of the area to be demolished by the Contractor in Zone 2A2 was handed over on 17 December 2007. The water line was not disconnected at the start of the demolition. Area 2A2 was also not handed over as per specifications, as construction debris and underground utilities were not removed (by others), thus causing the Contractor additional and abortive works resulting in disruptions and delays to his programme of work.

List of correspondence: GM/L-0531-07 on 24 October 2007; PMC-398-0747-07 on 17 November 2007; GM/L-0012-08 on 8 January 2008; PMC-398-0063-08 on 9 February 2008; GPM/L-0085-08 on 14 February 2008; PMC-398-0120-08 on 18 February 2008; GM/L-0108-08 on 24 February 2009; PMC-398-0144-08 on 27 February 2009.

### 6.4.5 Delayed handing over of Zone 1F

The Contractor's external works were hampered by the slow handover of Zone 1F by the Infrastructure Contractor. The Employer confirmed that the demolition of the gate house, access road and fence, and other utilities involved in Zone 1F were included under the contract scope of the main infrastructure package. The Employer added that the Contractor would be advised as soon as the area was ready for handover to enable the Contractor to start its works. The Contractor submitted the following proposal to mitigate delays.

1. The Employer to directly appoint a subcontractor to construct the permanent fence.

2. The Employer to directly appoint a subcontractor to erect a temporary fence between Area (A&C) and Area B.
3. The Contractor to demolish the existing fence between the main road and Area A and complete the required earthwork in Area A.
4. When the permanent fence is complete, the Contractor will demolish the existing fence between the main road and Area B, and complete all necessary earthworks.

List of correspondence: GM/L-0048-06 on 9 March 2006; PMC-398-0091-06 on 25 March 2006; PM/L-0041–08 on 26 January 2008; PMC-398-0287-09 on 28 April 2009; PM/L-0212-09 on 1 May 2009.

## 6.5. Delay analysis

The contract stipulated a specific timeline (start and completion) for each zone in the project in order to allow the Contractor to perform the works in a planned and orderly manner. The most dramatic loss of production, experienced by the Contractor, was the disruption by the Employer's directly employed Infrastructure Contractor's continuous delay in handing over the different zones in the project, which affected the mobilisation of crews and relocation of assigned resources, and which resulted in the Contractor needing to make efforts to deal with such disruptive and unknown conditions on virtually a daily basis to accommodate the site-availability needs instead.

Therefore, the Contractor, in this instance, is establishing entitlement by insisting on an extension of time and for an entitlement to payment of the losses caused by the delays and disruptions which arose because of the defaults of the Employer and/or his site representatives and/or his directly appointed Infrastructure Contractor.

For the purpose of this claim, the Contractor has established a right to an extension of time. The overall concept of causation is easy to grasp and established in this case, and is well documented in this claim report and supported in the attached Appendices. (Appendices are not included for the purpose of this case study. However, in a well-prepared claim, appendices must be well documented to support facts and numeration in the claim.) The Contractor, due to the excess and frequency of delays and disruptions in different stages and sections of Project A, is contending that all costs which they have incurred in excess of those anticipated at time of tender are to be attributed to the delays and disruptions to the different phases and zones.

In terms of delay, the Contractor has used a Time Impact Analysis approach.

This retrospective schedule analysis focuses on the impacted as-planned schedule to identify and quantify delays to the critical and near-critical paths of the project. This process involves the organisation and analysis of the extensive project documentation to establish the cause-and-effect relationships of each party's actions or inactions and the resulting delays.

Under this claim, the event (new duration, relation, milestone, etc.) is added to the schedule and the program is re-run to determine the impact of such delays on the project completion date. The difference between the original completion date and the new impacted completion date is the period of delay that occurred due to the particular delay event in question. This approach provides an objective assessment of the effect of the delay event when considered against the original critical path of the project.

The impacted delay analysis is performed using the schedules in hand to evaluate the contemporaneous, cumulative impact of delays caused by the Employer on the contractual completion dates. The delays to each activity affected are typically quantified by establishing the additional time required to perform the changed work or by including any time extension agreed in the approved change order in the critical path leading to project completion.

The Contractor prepared and submitted, in accordance with the conditions of contract, a plan of work (a Baseline Programme) based on the information and documents made available in the contract. This baseline programme considered a particular plan, resource allocation methodology and sequence of operation in order to meet the contractual completion dates of Project A.

As explained in previous sections, the delays incurred were due to causes beyond the control and liability of the Contractor. These delaying events have caused delays and disruptions that prevented the Contractor from progressing according to the planned schedule of works.

The detailed schedule analysis procedure includes

- preparation of duration and variance tables
- identification of delays and allocation of delay responsibility
- schedule calculations to determine delays.

Based on the retrospective analysis of the facts that had occurred on Project A, the Project activities experienced two types of delay.

- The first type of delay was a delay in the handing over dates. Consequently, new activities were added to the schedule to account for the delayed handing over.
- The second type of delay was related to the duration of activities that were disrupted, and consequently prolonged, due to causes beyond the control of the Contractor. For example, blocking of the deep water well, reallocation of the public toilet, etc. Consequently, new activities were added to the schedule to account for the new delayed work, or durations of existing activities were increased to show the effect of the increased work scope.

The impacts of the delays are tabulated in Table 6.1.

As set out in the attached claim documents, and in accordance with the conditions of contract referred to under this statement of claim, the Contractor hereby request the associated delay and disruption costs for events incurred up to 15 May 2010 (which

Table 6.1 Completion and impacted finish dates for Project A

| Activity description | Contractual finish date | Impacted finish dates | Difference in days |
|---|---|---|---|
| Completion base bid | 13 October 2008 | 10 January 2011 | 819 |
| Area 1BC | 23 February 2007 | 9 September 2008 | 564 |
| Area 1D | 25 November 2006 | 25 November 2006 | 0 |
| Area 1E | 18 December 2006 | 9 November 2008 | 692 |
| Area 1F | 27 March 2008 | 10 January 2010 | 654 |
| Area 2A1 | 16 August 2008 | 9 November 2008 | 85 |
| Area 2A2 | 25 July 2008 | 26 October 2009 | 458 |
| Area 3A2 | 21 April 2008 | 21 April 2008 | 0 |

was the actual completion date of Project A) and for acceleration costs occurring from 15 May 2010 to 10 January 2011 (the impacted extended completion date of Project A).

## 6.6. Evaluation of prolongation costs

The aforementioned delays adversely affected the Contractor by causing delays to critical and non-critical activities, disruption to the works, idle resources, and additional direct and indirect overheads.

When the Contractor submitted the price for executing the works, the Contractor included specific amounts for head office overheads and site-establishment costs. The cost of these resources were carefully planned and calculated, based on the original contract period.

The additional costs resulting from the prolongation of the time for completion, and costs incurred to mitigate delays and acceleration, are calculated in the claim evaluation section of this statement of claim.

Also, the Contractor's financial and human resources were planned and assessed in accordance with its cash flow and manpower projections, which were prepared at tender stage. The costs relating to the Contractor's financial arrangements and resources structuring had to be financed during the full contract period. The full prolongation cost and outstanding financing costs due on the unpaid sums are recoverable costs under the contract.

The cost impact on the Contractor is two-fold:

1. Delay and disruption costs which relate to the period from the contractual completion date of 13 October 2008 to the actual completion date of 15 May 2010 (581 days delay).
2. Acceleration costs to compensate for the period from 10 January 2011 (impacted completion date) to the actual completion date of 15 May 2010. These costs relate to the added resources such as labour, equipment and overtime costs for personnel to complete earlier than expected (240 days acceleration).

The Contractor is aware that, under the provisions of the contract, the Contractor may be entitled to extension of time in respect to events such as exceptionally inclement weather, strikes and the like, but the award of an extension of time with respect to such delays does not entitle the Contractor to additional payment for loss or additional expense. However, for other delays, for which the Employer or one of its representatives, being consultant, architect or directly employed contractor or subcontractor, is either responsible or has caused the delay, the Contractor shall and will, if such delay has affected operations as in these circumstances, be entitled to receive an extension of time and recover related loss and/or expense.

Therefore, the Contractor in this instance is not establishing entitlement to payment by only insisting on an extension of time, but also for an entitlement to payment for the cause of delay and disruption, which in this instance is caused clearly by the Employer and its site representatives and appointed contractors, and which resulted in extensive costs due to the numerous disruptions and delays during the course of the works. The Contractor is also claiming damages in respect of the acceleration costs it incurred to finish the works earlier than would have been the case under normal working conditions due to the impact of the delaying events.

The Contractor, due to the excess and frequency of delay and disruption events in different stages and sections of Project A, is contending that all costs which it has incurred in excess of those anticipated at the time of tender are to be attributed to the delay and disruptions or other causes for which the Employer is responsible.

This delay and disruption claim is essentially used to calculate the monetary claim which follows on from a delay to the work at Project A as a whole and in segments to the different zones as shown in Table 6.2. In Table 6.2, 'early finish' dates are the contractual finish dates and 'late finish' dates are the impacted finish dates for the various zones of Project A.

**Table 6.2** Early finish (EF), late finish (LF) and delay periods per zone*

| Activity description | Contractual finish date (EF): **1** | Impacted finish date (LF): **2** | Date of calculation of losses: **3** | Difference in days: **1–3** |
|---|---|---|---|---|
| Completion base bid | 13 October 2008 | 10 January 2011 | 15 May 2010 | 581 |
| Area 1BC | 23 February 2007 | 9 September 2008 | 9 September 2008 | 564 |
| Area 1D | 25 November 2006 | 25 November 2006 | 15 May 2010 | Nil |
| Area 1E | 18 December 2006 | 9 November 2008 | 9 November 2008 | 692 |
| Area 1F | 27 March 2008 | 10 January 2010 | 15 May 2010 | 654 |
| Area 2A1 | 16 August 2008 | 9 November 2008 | 9 November 2008 | 85 |
| Area 2A2 | 25 July 2008 | 26 October 2009 | 26 October 2009 | 458 |
| Area 3A2 | 21 April 2008 | 21 April 2008 | 15 May 2010 | Nil |

* Calculation of losses is made to the completion date of the project, which is 15 May 2010.

The losses are calculated based on the actual completion date of the project of 15 May 2010 (or an earlier date for a zone if that zone is completed prior to 15 May 2010). The acceleration costs have been calculated based on the anticipated impacted completion date of 10 January 2011, if the Contractor has not committed itself to mitigate the delays. Therefore, the period from 15 May 2010 to 10 January 2011 (240 working-days based on 7 days working per week) is the time by which the Contractor finished the works earlier than it would have done if acceleration measures had not been undertaken. The Contractor, in this claim, is seeking compensation for the acceleration costs he has incurred which he has otherwise not committed himself to.

If the impacted finish date is later than 15 May 2010 (which is the actual completion date) then the delay and disruption costs are based on the period up to 15 May 2010. The methodology used to calculate the damages is as follows.

- The Delay and Disruption Claim period relates the monetary amount related to the time difference between the contractual finish date (EF) and the impacted finish date (LF).
- The Acceleration Claim relates to the losses the contractor incurred to accelerate his works and to finish on 15 May 2010 instead of 10 January 2011, the date he would have finished but for his acceleration measures.

Therefore, the Contractor is entitled to additional time and costs insofar as those additional time and costs has been caused by the delayed start of each zone of Project A and the prolongation period as has been caused by the disruption caused by the Employer. The Contractor is also entitled to the acceleration costs as they have been incurred to the benefit of the Employer.

In this instance, and appropriately relating to each zone of Project A, where an extension of time is due because of conditions or obstructions that could not reasonably have been foreseen by an experienced contractor, the amount of any costs that may reasonably have been incurred by the Contractor by reason of such conditions or obstructions, together with a reasonable percentage addition thereto in respect of profit, and any extension of time to which the Contractor may be entitled, are the Contractor's lawful right and must be compensated by the Employer.

The starting point for the assessment of this delay and disruption claim to determine extension of time and losses is to establish the period for each zone for which the Contractor is entitled to further payment, and to apply an averaged figure for the Contractor daily prolongation costs, called 'delay and disruption costs'. Acceleration costs are calculated by applying the modified total cost method, which involves dividing up the additional costs and only claiming the certain parts of those costs that are the Employer's responsibility.

This claim focuses on the costs that are commonly claimed by contractors.

- *Unabsorbed overhead costs.* These costs include the site overhead indirect costs and head office overhead costs. These extra costs are directly related to this specific project. The site overhead indirect costs were tracked and quantified over the course of the alleged delay.

- *The site overhead indirect costs.* These costs include site supervision and administration, site security, office set-up, rental and similar expenses. The head office overhead costs are attributable to this specific project, and include items such as site management, site security, accounting, information systems, general insurance and indemnity, and financial adjustments such as bank overcharges and payment processing.
- *Profit.* The Contractor estimated profit is calculated as a percentage of the cost of each item in the bid tender package. The Contractor should be recovering profit during the delay and disruption period, and the key is to put the Contractor in as good a position as it would have been but for the Employer and his representatives' defaults or lack of actions delaying the project, and therefore the allowance of profit for the delayed period will provide the Contractor with a reasonable, just return.
- *Loss of labour productivity.* The loss of labour productivity is due to the fact that the Contractor could not execute the Project as planned and as it had priced the Project in the original tender bid. (These costs did not include the cost of overtime and the costs of lodging, transportation and logistics of keeping the labour for an extra 581 days, as these costs were absorbed in the overhead costs.) Costs or loss and expense in the nature of uneconomic working or loss of productivity of labour have been causing great hardship to the Contractor's operations and productivity through the duration of the project, due to works not being executed in their planned and logical sequences. Labour resources are working uneconomically, and additional cost has been incurred in order to maintain a required level of output and rate of progress.
- *Cost due to equipment and plant loss.* The extra equipment costs are based on the idle period life of each zone and include the costs of keeping the equipment in operational form. These costs include the loss due to management conditions. While operators may not be able to operate, they take several steps to improve management conditions, including greasing and lubricating the equipment frequently, checking parts regularly, especially parts that are worn easily, and keeping the equipment and plant clean and protected due to the harsh weather conditions on site. The leasing costs are incurred whether the equipment is externally rented or internally brought in. The Contractor is fully aware that loss of productivity of equipment has been minimised by moving equipment to other zones, and this loss has been calculated by taking into consideration this effect. Other fixed equipment costs, such as generators, fixed tower cranes and temporary site power, have been calculated.
- *Loss of chance.* The Contractor was continually trying to mitigate and to make reasonable efforts to minimise the costs it sustained as a result of the delays which were totally outside its control. The costs that need to be awarded to the Contractor include the loss of chance or opportunity due to the disruptions, financial costs such as interest paid, bonds and warrantees, bank charges, accounting, etc., and loss of productivity of labour, equipment and plant, and subcontractors' increased costs, which were also caused by the aforementioned disruptions.

This complex construction project gave rise to many, if not all, of these claim components in the execution of a single construction project. In addition to the quantification of costs, project records are kept by the Contractor in order to prove liability. Contemporaneous memoranda and related documents are also available to help support issues related to the causation of delay and other claims. Common information that the Contractor prepared and maintained includes

- job cost reports
- daily logs and progress reports
- daily payroll records
- change orders and related support
- meeting minutes
- job correspondence
- documentation of design changes
- photographs.

The Contractor gathered and maintained information to track the time, materials and other costs incurred on each phase and zone of the Project. This enabled the preparation of this statement of claim by assessing what work was performed, how much time it took and how much it cost. This information was used to illustrate what went well on the project before a delay or disruption event occurred, and the subsequent loss in productivity, idle time and damages after the numerous delay and disruption events occurred.

For the acceleration part of this claim, the modified total cost approach is used for the Project. The modified total cost approach addresses the potential problems associated with the total cost approach, where the claim amount is based on the difference between the contract actual cost and the bid cost, by adjusting or modifying for certain factors. For example, costs related to change orders and other claim items that do not relate directly to the delays in this claim are adjusted out of the claim calculation. Likewise, costs incurred by the Contractor that are not recoverable from the Employer, such as materials costs, specific subcontractor costs, and works to areas not affected by the delay even though works have been disrupted in such areas due to the interaction of activities of the whole project, are eliminated from the statement of claim. Also, any necessary modifications to the costs estimated in the bid, such as deductions and adjustments in the original tender, are taken into consideration.

The methodology applied has taken into consideration

1. segmenting costs to the delayed project as per delayed zone
2. using the same percentage of each type of segment of cost as that included in the bid
3. applying the different segments to calculate disruption and delay extra costs
4. calculating the acceleration costs based on the modified total cost method by calculating only the related costs due to acceleration such as labour, equipment, and direct and indirect costs.

The calculation of the different increased costs for the extended period of the project is prepared using the bid mark-up for site indirect overhead, head office overhead, profit, labour, equipment and subcontractors costs. The bid mark-up is the actual percentage by which each item in the tender is divided. For this project the bid mark-up is divided as follows.

1. Labour costs, which is labour direct cost and loss of productivity = 14% of the bid.
2. Equipment which is the cost of leasing, renting and purchasing equipment and plant for this project only. This cost includes salvage value and the cost of operation and maintaining the equipment = 6.2%.
3. Direct costs or head office overhead costs = 3% of the bid.
4. Indirect costs or site indirect overhead = 12% of the bid.
5. Profit = 12% of the bid (this percentage relates to Project A and is obtained from the original Bill of Quantities).

The formula applied is detailed hereunder:

Formula:

Daily rate for each claimable item × delayed period = compensation

The delayed period = LF − EF (where LF = late finish and EF = early finish)

As described above, early finish (EF) is the contractual finish date and late finish (LF) is the impacted finish dates. If the impacted finish date is later than 15 May 2010 (which is the actual completion date) then the delay and disruption losses are based on the period up to 15 May 2010.

The contract sum is $259 200 000.

### 6.6.1 Delay costs calculation

**Site indirect overhead**

Daily overhead, including office overhead, is 12%. Therefore, the claimable amount due to overhead losses = (12% × $259 200 000) divided by 1227 days = $25 349 per day.

Prolongation cost due to overhead for each zone = $25 349 × percentage of zone value × number of days delayed per each zone.

**Profit**

Daily profit as provisioned in the tender bid is 12%. Therefore, the claimable amount due to overhead losses = (12% × $259 200 000) divided by 1227 days = $25 349 per day.

Prolongation cost due to profit for each zone = $25 349 × percentage of zone value × number of days delayed per each zone.

**Head office overhead costs**

Daily office costs and as provisioned in the tender bid at 3%. Therefore, the claimable amount due to overhead losses = (3% × $259 200 000) divided by 1227 days = $6337 per day.

Table 6.3 Value and delayed period per zone

| Zone | Construction value: $ | Percentage of zone value: $X$ | Early finish: EF | Late finish: LF | LF–EF: days ($Y$) |
|---|---|---|---|---|---|
| 1BC | 44 143 126 | 17% | 23 February 2007 | 9 September 2008 | 564 |
| 1D | 8 523 715 | 3% | 25 November 2006 | 25 November 2006 | 0 |
| 1E | 49 847 691 | 19% | 18 December 2006 | 9 November 2008 | 692 |
| 1F | 4 504 133 | 3% | 27 March 2008 | 10 January 2008 | 654 |
| 2A1 | 31 900 750 | 12% | 16 August 2008 | 9 November 2008 | 85 |
| 2A2 | 80 114 541 | 31% | 25 July 2008 | 26 October 2009 | 458 |
| 3A2 | 37 752 610 | 15% | 21 April 2008 | 21 April 2008 | 0 |

Prolongation cost due to head office overhead for each zone = \$6337 × percentage of zone value × number of days delayed per each zone.

**Labour direct cost and loss of productivity**

The average labour cost for the Project is 14.1%. Therefore, the daily average labour direct costs = (14.1% × \$259 200 000) divided by 1227 days = \$29 785 per day.

Prolongation cost due to labour for each zone = \$29 785 × percentage of zone value × number of days delayed per each zone.

**Equipment and plant**

The average cost for the equipment and plants is, on average, 6.2% for the whole project. Therefore, the daily claimable amount due to plant and equipment idle time = (6.2% × \$259 200 000) divided by 1227 days = \$13 097 per day.

Table 6.4 Site indirect overhead, head office overhead and profit

| Zone | Site indirect overhead: $ | Head office overhead: $ | Profit: $ |
|---|---|---|---|
| | 12% × \$259 200 000/1227 = \$25 349 | 3% × \$259 200 000/1227 = \$6337 | 12% × \$259 200 000/1227 = \$25 349 |
| | \$25 349 × $X$ × $Y$ ($X$ and $Y$ are from Table 6.3) | \$6337 × $X$ × $Y$ ($X$ and $Y$ are from Table 6.3) | \$25 349 × $X$ × $Y$ ($X$ and $Y$ are from Table 6.3) |
| 1BC | 2 430 460 | 607 590 | 2 430 460 |
| 1D | 0 | 0 | 0 |
| 1E | 3 332 880 | 833 180 | 3 332 880 |
| 1F | 497 340 | 124 330 | 497 340 |
| 2A1 | 258 550 | 64 630 | 258 550 |
| 2A2 | 3 599 050 | 899 720 | 3 599 050 |
| 3A2 | 0 | 0 | 0 |
| **Total** | **10 118 280** | **2 529 450** | **10 118 280** |

Table 6.5 Labour direct cost, loss of productivity and equipment and plant costs

| Zone | Labour direct cost and loss of productivity: $ | Equipment and plant: $ |
|---|---|---|
| | 14.1% × $259 200 000/1227 = $29 785 | 6.2% × $259 200 000/1227 = $13 097 |
| | $29 785 × *X* × *Y*<br>(*X* and *Y* are from Table 6.3) | $13 097 × *X* × *Y*<br>(*X* and *Y* are from Table 6.3) |
| 1BC | 2 855 780 | 1 255 740 |
| 1D | 0 | 0 |
| 1E | 3 916 130 | 1 721 990 |
| 1F | 584 380 | 256 960 |
| 2A1 | 303 800 | 133 580 |
| 2A2 | 4 228 870 | 1 859 510 |
| 3A2 | 0 | 0 |
| **Total** | **11 888 960** | **5 227 780** |

Prolongation cost due to equipment and plant for each zone = $13 097 × percentage of zone value × number of days delayed per each zone.

Tables 6.3, 6.4 and 6.5 itemise the percentages of the construction cost of each zone as a proportion to the total cost, and the early finish and late finish for each zone. The construction amount for each zone was calculated as the percentage each zone constitutes in proportion to the whole Project lump sum price. For example, the construction value of Zone 1BC in Table 6.3 is equal to the project value ($259 200 000) multiplied by 17% (the proportion of Zone 1BC in relation to the whole Project), which is $44 143 126. The tables also show the claimable amount for each type of damage and for each zone.

Table 6.6 summarises the total delay cost, which amounts to $39 882 750.

Table 6.6 Summary of the total delay cost, which amounts to $39 882 750

| Delay and disruption cost distribution | Total cost: $ |
|---|---|
| Site indirect overhead (from Table 6.4) | 10 118 280 |
| Head office overhead (from Table 6.4) | 2 529 450 |
| Profit (from Table 6.4) | 10 118 280 |
| Labour direct cost and loss of productivity (from Table 6.5) | 11 888 960 |
| Equipment and plant (from Table 6.5) | 5 227 780 |
| **Total delay and disruption costs** | **39 882 750** |

Table 6.7 Summary of the acceleration costs, which amount to $17 896 320

| Acceleration cost distribution | Total cost: $ |
|---|---|
| Site indirect overhead | 6 083 760 |
| Head office overhead | 1 520 880 |
| Labour direct cost and loss of productivity | 7 148 400 |
| Equipment and plant | 3 143 280 |
| **Total acceleration costs** | **17 896 320** |

### 6.6.2 Acceleration costs calculation

Acceleration is due for the total amount of 240 days. The costs associated with acceleration are summarised in Table 6.7 and detailed below.

**Site indirect overhead**

Daily overhead including office overhead is 12%. Therefore, the claimable amount due to overhead losses = (12% × $259 200 000) divided by 1227 days = $25 349 per day.

Acceleration cost due to site, indirect, for the whole project = $25 349 × 240 (the number of days accelerated) = $6 083 760.

**Head office overhead costs**

Daily office costs as provisioned in the tender bid at 3%. Therefore, the claimable amount due to overhead losses = (3% × $259 200 000) divided by 1227 days = $6337 per day.

Acceleration cost due to head office overhead for the whole project = $6337 × 240 (the number of days accelerated) = $1 520 880.

**Labour direct cost and loss of productivity**

The average labour cost for the Project is 14.1%. Therefore, the daily average labour direct costs = (14.1% × $259 200 000) divided by 1227 days = $29 785 per day.

Acceleration cost due to labour cost for the whole project = $29 785 × 240 (the number of days accelerated) = $7 148 400.

**Equipment and plant**

The average cost for the equipment and plants is, on average, 6.2% for the whole project. Therefore, the daily claimable amount due to plant and equipment idle time = (6.2% × $259 200 000) divided by 1227 days = $6337 per day.

Acceleration cost due to overhead for the whole project = $13 097 × 240 (the number of days accelerated) = $3 143 280.

## 6.7. Total amount of claim

| | |
|---|---|
| Delay and disruption amount | $39 882 750 |
| Acceleration amount | $17 896 320 |
| Total | $57 779 070 |

**Delay and Disruption Claims in Construction**
ISBN: 978-0-7277-5705-0


doi: 10.1680/ddcc.57050.085

# Chapter 7
# Case study: loss and expense, disruption claim (including a global claim example)

This case study seeks to demonstrate an entitlement to loss and expense in a 'real-life' (although fictitious) situation, and the format (although obviously not the content) of this case study can be used as the basis of any such claim. This claim is classified as a modified total cost claim, where the costing methodology is different from a typical loss and expense claim, and where the claimant claims only for the costs that cannot be quantified on an individual basis, and the amounts claimed relate to delays and disruptions caused entirely and solely by the client.

It is useful to start any claim submission with an executive summary. The executive summary provides an overview of the entire claim, and makes it simple for anyone reading the claim to understand the basis of the claim and what is being sought as redress from the claim.

A background to the claim submission then needs to be provided, followed by a narrative of events. A legal analysis to show the basis of the claim should then be given, followed by a detailed costing methodology.

In line with the above comments, the case study below is set out under the following headings

- executive summary
- background
- narrative of events
- legal analysis
- costing methodology.

## 7.1. Executive summary

This claim includes the legal and contractual background on which the Contractor is basing its right for compensation, and which has imposed a heavy burden on the operation of the project and placed the Contractor under extreme economic duress due to the heavy losses that it has incurred. Project A was marred by numerous problems from the beginning, when the Contractor mobilised his staff, workforce and resources. The number and impact of these problems are rarely encountered in one single project, and if only one or two of these problems occur on any one particular project they can inflict heavy losses on the Contractor.

Each claim heading is written in detail, and all the legal background, the problems encountered on the project and the amounts to be claimed are supported with facts and documents, including letters, reports, invoices, tender documents, subcontractors' correspondence and variation orders. The Contractor is seeking remedies to the problems encountered in the Project, and relief though common and applicable laws by applying certain doctrines such as misrepresentation, economic duress and frustration, which allow a degree of release from the terms of the contract or adjustment of these terms in the face of the exceptional unforeseen conditions that occurred through the life of the project.

The delays and obstacles that occurred could not be foreseeable by the Contractor or any other experienced contractor, and were outside the scope of the assumed risks considered under the contract at the time the job was awarded on 3 November 2006 and during the period prior to that date when the project was estimated and the plans to construct were put in place. The consequence of the above disruptions was such that the contract could have been rescinded due to the doctrine of frustration where, in the event of frustration, the contractor will be discharged in relation to future performance obligations. The Contractor was aware of its right to rescind the contract due to these circumstances but decided to continue the work, even though it was incurring heavy losses, amounting to more than $4 000 000, and even though the project would have been delayed beyond any reasonable date that could have been envisaged and foreseen if it had not been for the Contractor's acceleration measures.

The Contractor faced enormous difficulties from the outset and these are explained in detail in the claim document. The Contractor contends within this statement of claim that the changes and disruptions that occurred during the life of this project amount to a fundamental change in the scope of the work and that the additional costs were clearly unforeseeable. The Contractor was denied the costs of the cumulative impact of the above, and it can rightfully claim, and it is clearly entitled to, additional compensation as a result of the numerous change orders that had cumulative impacts throughout the execution of the project. Notwithstanding the lack of absolute direct causal proof for individual events, it is clear – on the balance of probabilities – that by way of the 'but-for' test, if it had not been for

- the employer's numerous, pervasive and cumulative changes instructed throughout the execution of the works
- the employer's onerous security procedures and bureaucratic requirements
- the remoteness and harshness (social and environmental factors) of the site
- the consultant's inefficiency and lack of necessary staff.

The Contractor would have been able to complete the works within budget. However, the employer caused major delays and disruption and, as such, and as detailed in this loss and expense claim, the Contractor is entitled to compensation for cumulative disruption of $4 268 429.

## 7.2. Background

The parties to the contract to which this claim is subjected are as follows.

1. International Operations Client Limited (the Client)
2. Prime Contractors Limited (the Contractor).

Project A is an extensive construction project with a value of $31 987 186 and was being constructed in three phases over a construction period of 1396 days. This claim is prepared after the completion date of the project, and the Contractor has submitted all necessary notices in due time to preserve his right to submit this claim.

This is a loss and expense claim with a value of $4 268 429 relating to Project A. The Contractor entered into contract with the Client on the 3 November 2006, to construct the overall development. The contract commenced on the 28 November 2006 and was due for completion 1396 days later on 28 October 2010.

The Contractor faced enormous difficulties from the outset, and these are explained in great detail in this claim document. However, synopses of these difficulties are summarised in the following parts of this claim.

## 7.3. Narrative of events

### 7.3.1 Disrupted start

An extended period of time (almost a year) was taken from the initial contract award date (the date of the letter of intent) up to the actual date of signing the contract. Such a prolonged idle period forced the Contractor to reallocate the pre-assigned project team to other ongoing projects, a matter that had a negative impact on the project mobilisation and associated timely start of the work. This also has resulted in the Contractor releasing some of its workforce or distributing some of its workforce to other projects.

### 7.3.2 Disruption due to piling works

Based on the project records, the Contractor was facing great difficulties in soliciting specialist subcontractors for the piling works at the project, due to the fact that the drawings and specifications were incomplete for the foundations. Thus, the progress of piling works was adversely affected, despite all positive attempts by the Contractor. In addition, the soil conditions data provided in the project geotechnical report were not compatible with the actual site conditions, and this in turn delayed the start of the substructure works.

### 7.3.3 Impact of project location on manpower recruitment

Since the start of Project A, the Contractor and its subcontractors have faced serious difficulties in attracting and recruiting qualified manpower due to the remote location of the project and the major boom in the construction field in the region where the project is located. The direct and indirect costs became more significant as a result of the knock-on effect on the costs of out-of-sequence work and considerable rework caused by the disrupted labour availability and the ever increasing cost of obtaining willing labour to work in the present environment considering the location of the site. There was an unprecedented movement of labour in and out of Project A and this severely disrupted the required level of output. Consequently, the difficulty in securing sufficient resources to satisfy the requirements of the planned activities severely degraded the Contractor's planned output and progress of work. Such scarcity of qualified manpower resources (which is a compensation event under the contract) remained a major concern to the Contractor until the completion of the works.

### 7.3.4 Client organisation and structure

The Contractor had to deal and report to different Client departments, sometimes for a simple process such as entering equipment on site. Aside from being located in a remote area, the project is within the regional strategic premises. Entry and exit to Project A is highly restricted, a matter that has hindered the movement of manpower, materials and equipment. Notwithstanding the Client binding safety and security rules, works at site have been placed on hold on several occasions. This is attributed to the strict regulations and their implementation by the concerned safety officers, and the background and capability of the Contractor's manpower to abide the stringent safety rules and regulations, thus imposing additional burden on the Contractor.

### 7.3.5 Subcontractors' performance

Notwithstanding the fact that the project works are the prime responsibility of the Contractor, the Contractor was required to sublet part of his responsibilities to designated subcontractors.

For various reasons, one of which was the boom that the region experienced between 2006 and 2008, and due to the various delays, which had a great influence on the increase in cost inflation, the Contractor faced serious difficulties in appointing qualified and approved subcontractors. Also, due to the location of the site, and the heavy restrictions imposed by the Client with regard to access to the site, it was very difficult to appoint subcontractors or, if accepted, their prices were extortionate. Nevertheless, the Contractor has exerted all possible efforts to overcome such limitations through the deployment of key personnel. The Contractor's supportive attitude has alleviated the impact of such deficiency on the project's overall delays, but has put a heavy burden on his finances due the increased direct and indirect costs associated with this task.

### 7.3.6 Shop drawings and material submittals

The Contractor assumed that a substantial part of the contract drawings were well detailed and would be issued for construction based on normal practice. However, that assumption proved to be incorrect, particularly in the structural and architectural disciplines. In fact, over 2800 shop drawings needed to be produced by the Contractor simply to finalise design details that were not the Contractor's liability.

Conceptual design of various items that required detailed designs led to full reviews by the Engineer. Such reviews were time consuming, specifically due to a shortage of the Engineer's staff – in fact the Engineer was not fully staffed until November 2007.

As a result of the above, 70% of the shop drawing submittals were subject to extended review periods, which were beyond the allotted contract float and which ranged from 1 day to over 75 days, as set out on the Contractor's shop drawing submittals log.

Similarly, 37% of the submittals were returned to the Contractor beyond the allotted review period, ranging from 1 day to over 57 days, again based on the Contractor's relevant submittals log.

Although providing the Contractor with an electronic version of contract drawings was a contractual obligation on the part of the Client, several months elapsed before an electronic copy was provided to the Contractor. Thus, the time period for the preparation of shop drawings and submission was considerably extended.

### 7.3.7 Late arrival of the consultant's team to site

It is imperative that projects of the size and complexity of Project A are given early momentum in order to achieve a rate of output that will result in the execution and completion of the works in accordance with the planned schedule envisaged during the tender process. Essential to this aim is the early and proactive involvement of the consultants. Unfortunately, on Project A the arrival of the consultant's site-based teams, in the numbers required to facilitate the necessary early momentum, was very slow, and this had a dramatic negative effect on the Contractor's planned rate of output in the early months. A proactive consultant approval of the Contractor drawing and material submittals is essential to facilitate the Contractor's need to gain this early momentum. Unfortunately, the consultants failed to provide the early assistance required. Apart from the drawings and materials approvals, which are essential in the early phase of any project of this size and complexity, the consultant teams were needed in sufficient numbers to address the many queries the Contractor formulated in the early phases of the works. In addition to addressing the many queries, the consultants were required to issue many drawing clarifications to ensure that the Contractor could commence early construction activities in a meaningful and purposeful manner.

### 7.3.8 Disruption arising from change orders and disputed change orders

A considerable volume of change orders was issued during the construction period, and these changes, combined with the very considerable number of disputed change orders, as detailed in the body of this claim document, caused considerable cumulative disruption to the Contractor's regular progress of work. The Contractor asserts that the number, timing and effect of the changes issued impacted his ability to plan and perform the works. The existence of unplanned events and conditions is proved beyond any reasonable doubt. The Client, by issuing this great number of change instructions, disrupted the works immensely. The cumulative impact of the many Client-initiated changes over the duration of the works, which was beyond the Contractor's control and foreseeability, is clearly the most probable cause of the delay and the ripple disruptive effects that have impacted the disruption, and the great losses the Contractor has suffered during the execution of Project A.

Multiple change orders have been issued on Project A; and there is a solid *prima facie* case showing that the Client's numerous change orders form a basis for the Contractor to globally claim, as it is impossible to differentiate between their cost impacts or to calculate some of the change orders individually due to their complex interdependency. Where it was possible to do so, the Contractor has highlighted the cost of impact of the separate variations to be deducted from the total loss (a process known as the 'modified total cost method').

The Contractor asserts that

- the Client's disruptions caused a material change in the nature, scope and schedule of the project, and the Contractor suffered the effects of the events, a cumulative impact on productivity and heavy losses due to the mentioned disruptions
- the changes amount to a fundamental change in the scope of the work, and the additional costs were unforeseeable.

The Contractor was denied the costs of the cumulative impact which any other contractor can rightfully claim if subjected to the same conditions as faced in Project A. As such, the Contractor is clearly entitled to additional compensation as a result of numerous change orders and the other mentioned reasons that had cumulative impacts throughout the execution of the Project.

## 7.4. Legal analysis

Project A was marred by delays from its starting date, and was subjected to many unforeseeable factors, including design errors and omissions, a large number of Client-initiated changes and variations, unanticipated obstacles due to site restrictions, restrictive bureaucratic practices and unforeseen lack of resources.

The Contractor is relying on its awareness of the general policy of the law towards commercial transactions, including construction contracts, which is based on freedom of contract within an established framework of rules. But the Contractor would like to emphasise that this is balanced by countervailing policies, some of which might be explained on the grounds of either public policy or doing justice.

In this claim, the Contractor is making demands for loss and expense as reasonable payment to compensate for damages that accrued in this project. The courts have intervened by imposing liability in the absence of, or alongside, contractually binding obligations, and the Contractor hereby is relying on the stated legal doctrines for relief against the binding contractual obligations that has forced him to work under extreme duress.

In this claim, the Contractor stresses that there are other areas where the law intervenes to provide relief or remedies, despite apparently binding agreement or in the absence of binding agreement, such is the case in this particular contract drawn between the Client and the Contractor. Where the courts intervene to upset apparently binding agreements, they do so to preserve the rights of the parties that had endured economic duress, such as in the case of the Contractor in this particular project. Freedom of contract is, of course, the paramount contractual policy, and most of the disruptions and duress that the Contractor was subjected to are deviations on that principle.

The point made by J. Jessel in *Printing and Numeric Registering* v. *Sampson* (1875) LR 19 EQ 462 at p. 465, to a plea that a contract should be held unenforceable on the grounds of Economic Duress, is equally valid in other areas which involve the courts in declaring a contract void, voidable, unenforceable or discharged before performance is completed:

> It must not be forgotten that you are not to extend arbitrarily those rules which say that a given contract is void, because if there is one thing which more than another policy requires, it is that men of full age and competent understanding shall have the utmost liberty of contracting, and that their contracts when entered into freely and voluntarily shall be held sacred and shall be enforced by courts of justice. Therefore you have this paramount policy to consider – that you are not lightly to interfere with freedom of contract.

The doctrine was recognised by the Privy Council in *Pao On* v. *Lau Yiu Long* [1980] AC 614, where the possibility of economic duress as grounds for setting aside the contract was acknowledged. It was also recognised and applied by the House of Lords *Dimskal Shipping Co. SA* v. *International Transport Workers' Federation, The 'Evia Luck'* [1991] 4 All ER 871:

> A contract entered into under duress is voidable and void. A person who has entered into the contract may either affirm or avoid such contract after the duress has ceased; and if he has so voluntarily acted under it with the full knowledge of all the circumstances he may be held bound on the ground of ratification, or if, after escaping from the duress, he takes no steps to set aside the transaction he may be found to have affirmed.

Therefore, and understanding that this contract has been entered into by the parties freely, the Contractor has done its utmost not to declare this contract void, voidable or unenforceable, even though it has been working under tremendous duress and incurring losses that other contractors almost certainly could not sustain. This fact and others could render any type of contract under the ambit of law and courts as dischargeable.

### 7.4.1 Applicable laws related to this claim

The above factors that hindered the progress of the works and caused tremendous damage to the Contractor during the course of the project, allow a degree of release of the terms of the contract or adjustment of these terms in the face of the exceptional unforeseen conditions that occurred through the life of the project until the present. The Contractor's failure to comply with the terms of the contract, or even accepting the penalties imposed in rescinding the contract, would have meant that its liquidated damages liability would amount to no more than $3 198 718. The Contractor was working under huge, ongoing losses that amounted to $4 268 429, which is beyond any reasonable amount that should or could be incurred under the contract, and reasons for delays, which have effectively made the time for Project A 'at large'.

This case, therefore, raises sharply the question of what is the nature and extent of the duty of the Client whose contract operations have caused heavy losses to the Contractor executing the works. The Contractor in this instance asserts that the Client must not carry out or permit the continuation of an operation which he knows or ought to have known would clearly cause such losses, however improbable that result may be. The Contractor asserts that the Client is required to take into account the possibility of such damage, particularly if the damage is such that a reasonable Client, careful of the operations of the project being executed, would regard it as material.

The test to be applied here is whether the risk of damages to a contractor was so small that a reasonable client in the position of the existing Client, considering the matter from the point of view of profitability, would have thought it right to refrain from taking steps to prevent the losses. In considering the matter, the Client's right to take into account, not only how remote the chance is that the Contractor might lose so great an amount of money as to put his whole business in jeopardy, but also how serious the consequences are likely to be for the Contractor's losses, to take into account the difficulty of the remedial measures.

The delays, disruptions and obstacles that occurred could not be foreseeable by the Contractor or any other experienced contractor, and were outside the scope of the assumed risks considered under the contract at the time the project was awarded in October 2006 and during the period prior to that date when the job was estimated and the plans to construct were put in place. The consequence of the above disruptions is that the contract could be rescinded due to the doctrine of frustration, where in the event of frustration the contract can be discharged in relation to future performance obligations of both parties.

The Contractor, even though knowing its right to rescind the contract due to these circumstances, decided to continue the work, even though under heavy losses amounting to more than $4 million, and the project could have been delayed beyond any reasonable date that could have been envisaged and foreseeable if it had not been for the acceleration procedures carried out by the Contractor to complete within a reasonable time, considering the delays and disruptions caused by the Client.

### 7.4.2 Frustration of contract

The courts have been willing to provide relief against disadvantageous contracts, and they have done so by coordinated steps with common philosophy and principle. Where the law has intervened, the approach has generally been universal in providing relief and remedies to the injured party. Therefore, The Contractor is insisting that the contract, which it entered into in good faith, is disadvantageous, and the terms of the contract are not fair, even though negotiated in good faith. There is, in any event, the possibility of the parties to a contract incorporating a duty of good faith, a remedy when encountering unforeseen circumstances or when abrogating factors occur, such as misrepresentation, and this may be increasingly encountered in construction contracts, as in the present one.

Under various laws as well, frustration sometimes provides relief where the performance or purpose of the contract has been made impossible or illegal by supervening events (i.e. events taking place after the contract was formed), which were not foreseen at the time when the contract was formed, and which were not due to any act or omission of either party. The doctrine applies primarily to supervening events making performance impossible or illegal, and which therefore make the contractual obligations no longer binding on the parties. The contract in this instance did not make any allowance for the problems that the Contractor has experienced during the execution of Project A, such as the late staffing of the consultants' teams at the start of the Project, or the lack of initial design, which progressed into a large number of change orders. All the problems that were encountered could not be foreseen by either the Client or the Contractor, and could therefore be attributed to the doctrine of Common Mistake, where the heading 'mistake' refers to situations where one or both parties to a contract are under a misapprehension of present fact at the time of contract.

In summary to the aforesaid, the elements of the law that are pertinent to this particular loss and expense claim are misrepresentation, frustration and economic duress. The acts, preventions and large number of change orders provide the Contractor for Project A enough reasons under applicable laws to rescind this contract, and, if not so, to substantially succeed in recuperating his losses under this statement of claim.

Rescinding the contract early would have minimised the Contractor's exposure to loss, but it would have in turn exposed the Client to considerable additional costs. The project would have been re-tendered, or negotiated with the next lowest tenderer, and the Client would have faced considerable delays and additional costs and losses. However, the Contractor valued its long-standing relationship with the Client, and its impeccable record in the field of construction, and as a result did not pursue the rescission of the contract. It did so trusting that an amicable settlement would be reached with the Client when the works were nearing completion.

### 7.4.3 The global claim approach

This loss and expense claim falls under the ambit of global claims, where the Contractor does not seek to attribute loss to specific breaches of the contract or to specific legal factors and/or doctrines that rendered this contract unworkable, but rather alleges a composite loss as a result of all the notified breaches by the Client. A global claim is, in this instance, defined as 'the antithesis of a claim where the causal nexus between the wrongful act or omission of the defendant and the loss of the claimant has been clearly and intelligibly pleaded' (*Bernhard's Rugby Landscapes Ltd* v. *Stockley Park Consortium Ltd* (1997) 82 BLR 39).

This global claim is manifested by not reconstructing what actually happened on site, as it is impossible or impractical to plead and prove the causal nexus. The factors and reasons that the Contractor has incurred are so intertwined that he could not possibly identify how each breach has caused certain damage and, therefore, an amount of loss. Therefore, the contractor is pleading his case using a global claim, as global claims are only permissible when it is impossible or impractical to plead and prove the causal nexus (see *J. Crosby & Sons Ltd* v. *Portland Urban District Council* (1967) 5 BLR 121 and *London Borough of Merton* v. *Stanley Hugh Leach* (1985) 32 BLR 68).

## 7.5. Costing methodology

The composite loss in this claim is prepared using a modified total cost method (*Servidone Construction Corporation* v. *The United States* 931 F.2d 860. April 24, 1991), where the quantification of loss is achieved by subtracting the tender cost of the works from the final cost. Then, this total cost is further adjusted by subtracting from it the direct and indirect costs, and then subtracting all the variations that have a direct cost impact and applying the material escalation factor. The variations with a direct cost impact and the material escalation will be claimed separately under different heads of claim.

In this claim, the Contractor argues that the events which occurred during this contract period were complicated to such a degree to make it impracticable, if not impossible, to assess the additional expenses caused by delay and disorganisation due to any one of the events in isolation from the other events. While the Contractor may be able to provide a list of numerous events that have caused disruption to his works, the global claim does not prove what the effects of such disruptive events really were to the works, as they cannot be defined in terms of quantum. It can be argued that there can be many reasons why the Contractor's final costs are more than his tendered costs, as he may simply have tendered too low in the first place. In this case, the Contractor won the contract through a tender process while competing with a number of other bidders. Further negotiations were conducted between the Client and the Contractor after the tender process had ended, which resulted in a further reduction in his bid, and therefore the cost of his works cannot be argued as being too low.

The Client could argue that the Contractor's loss and expense claim makes the great assumption that all the additional time and costs were caused by the disruptive events that are included for the purpose of this claim, and that would be correct. The Contractor is claiming, justifiably, that all the events that caused the disruption to the works were the Client's responsibility, but in acting in a fair and reasonable manner the Contractor has deducted from his calculation all unrelated direct and indirect costs, as well as the profit margin. The loss the Contractor is claiming for in this claim is only the actual losses in respect of the works that he was required to carry out over and above its original scope of work. It is also to be noted that the Contractor has accelerated the works to finish them by the agreed completion date, as instructed by the Client.

The loss and expense claim pursued is based on an allegation that there were numerous variations to the contract, and the cost overran. The Contractor asserts that the cost overrun is recoverable as a result of the variations and the Client-initiated causative factors, such as lack of design, late approval of the design and late materials submittals. There is, however, no analysis that a particular causative factor led to a particular item of loss. This claim is based on the principle that there were numerous events interfering with the works, entitling the Contractor to recover damages for incurred losses. Again, there is no link between the alleged events and the resultant damages, but the Contractor is claiming that the disruptive effects, together with the legal deviation from the Contract, have contributed as a whole to a lump sum loss of \$4 268 429.

The law states that, if parts of the claim can be pleaded and proved on a conventional basis (i.e. direct cause and effect) and other parts cannot, only the latter can and should be presented on a global basis. In this case, the Contractor, in his modified total cost approach, has eliminated from his global claim all losses that could be claimed conventionally, such as

1. material escalation due to the unrealistic and unenvisaged increases that were recorded in the prices of all construction materials, mainly from 2006 to 2010
2. variations that have a direct cost impact and have an identified monetary value.

The Contractor realises that the correct manner of presenting a claim is to link the cause with the effect. However, in this case this is not achievable, especially when the particulars of this loss and expense claim, rather than a prolongation claim, are disruption events with no methodology or a systematic approach to allow their calculation. To counter such difficulties, the Contractor has proved that to link cause and effect is not possible, and thereby the use of this partially global claim is acceptable in the circumstances.

### 7.5.1 Loss and expense components

In the disruption associated with the cumulative changes and causes for the construction of Project A, it is clear that

1. The cumulative impact of the excessive changes and the disruptive causes in the project affected the performance and schedule of works significantly.
2. The cumulative impact of the excessive changes and disruptive causes increased the cost of performance.
3. The impact was not foreseeable when the disruptions occurred and the change orders were acknowledged.
4. The changes were so numerous and the disruptive causes were so overlapping that it constituted reasoning for this global claim with a modified approach.
5. The agreed changes were so numerous or overlapping that the Contractor obtained less than a full recovery for the individual changes when approved in isolation.
6. At the time when the Contractor negotiated the changes approved by the Client, he had no way of knowing the full impact of the changes and the disruptions on the Project overall final cost.

The enormous difficulties detailed above on Project A have resulted in the Contractor incurring extremely heavy losses. The Contractor submits that the Client is duty bound to compensate the Contractor for the extremely heavy losses incurred and the effort taken for the drawings approval and materials submittals approval on this project for reasons that were completely outside the Contractor's control and completely outside the Contractor's contemplation at the time of submitting the tender.

Furthermore, as is clearly documented in the rest of this claim document, the Contractor contends that he had the right to seek relief under applicable laws from the excessive losses described above. The elements of the law that are pertinent to this particular contract are

1. misrepresentation
2. frustration
3. economic duress.

The Contractor contends that the contract could have been rescinded under at least one, if not all three, of these legal concepts. Rescinding the contract for reasons of misrepresentation, *force majeure*, frustration or economic duress would have allowed the Contractor to limit its losses in the early stages of the project, and to have realised an overall loss of a fraction of the losses it incurred at the end of the project. Rescinding the contract early would have minimised the Contractor's exposure to loss, but it would have in turn exposed the Client to considerable additional costs. The Project would have been re-tendered, or negotiated with the next lowest Contractor, and the Client would have faced considerable delays and additional costs and losses.

However, the Contractor valued its long-standing relationship with the Client, and its impeccable record, and as a result it did not pursue the way in rescinding the contract. It did so trusting that an amicable settlement would be reached with the Client when the works were nearing completion.

### 7.5.2 Calculation methodology

Total cost of the project $= C_1 =$ \$41 230 550

Total amount tendered $= C_2 =$ \$31 987 186

Total losses $= C_3 = C_1 - C_2 =$ \$9 243 364

Direct costs, indirect costs and profits = 27% of the total losses

Total net losses (after deducting direct costs, indirect costs and profits) $= C_3 \times 73\%$
= \$9 243 364 × 0.73 = \$6 747 656

The contractor has adjusted his total net losses by subtracting from this claim the variations agreed to be paid by the Client, which amount to \$1 038 747, and the other claim that he will be submitting in relation to the materials cost increase, which will total to \$1 440 480.

Total net losses = \$6 747 656

Agreed variations = \$1 038 747

Material escalation claim = \$1 440 480 (This claim will be submitted separately.)

**Loss and expense claim amount = \$4 268 429** (i.e. \$6 747 656 less \$1 038 747 less \$1 440 480)

## Cases

*Bernhard's Rugby Landscapes Ltd v Stockley Park Consortium Ltd* (1997) 82 BLR 39
*Dimskal Shipping Co. SA v International Transport Workers' Federation, The 'Evia Luck'* [1991] 4 All ER 871

*J. Crosby & Sons Ltd* v. *Portland Urban District Council* (1967) 5 BLR 121
*London Borough of Merton* v. *Stanley Hugh Leach* (1985) 32 BLR 68
*Pao On* v. *Lau Yiu Long* [1980] AC 614
*Printing and Numeric Registering* v. *Sampson* (1875) LR 19 EQ 462
*Servidone Construction Corporation* v. *The United States* 931 F.2d 860. April 24, 1991

**Delay and Disruption Claims in Construction**
ISBN: 978-0-7277-5705-0


doi: 10.1680/ddcc.57050.099

# Chapter 8
# Case study: increased costs claim

## 8.1. Project contract details

Project B is an extensive construction project for the redevelopment of existing housing business and service quarters into housing family units, bachelor quarters, a range of modern recreational facilities, improved office space, workshop and storage facilities, undertaken in the United Arab Emirates. The Project value as originally tendered is $518 400 000 and was planned to be constructed in three phases with a total construction period of 1227 days. The development consists of approximately 153 000 square metres ($m^2$) gross internal floor space and 245 000 $m^2$ external developments.

The pertinent details extracted from the Contract are

- Contractor: Contractor Construction Limited (the Contractor)
- Client: International Client Limited (the Client)
- Commencement date: 4 June 2005
- Original completion date: 31 December 2009
- Type of contract: lump sum
- Period between interim payments: monthly
- Retention: 5%.

## 8.2. Case statement

The Contract entered into between Contractor and Client is a Fixed Price Lump Sum Agreement, whereby all rates for materials, equipment and labour are fixed throughout the planned construction period of the Project. The risks associated with price escalation and inflation generally rest with the Contractor during the period of the project, as set out in the Contract.

During the tendering period for the works, the Contractor analysed the inflationary trends for all materials utilised for the construction of the Project, both those manufactured locally and those imported from the international market. This analysis was based on a 5-year period prior to the tender date, and the mean results from that analysis were used to allocate the costs and the risk for material inflation for the envisaged construction period. The cost of each type of material to be used was then obtained and multiplied by the factor that represents the forecasted increase in materials prices for the period of the contract.

However, two significant unforeseen events distorted this risk allocation mechanism for inflationary forecasting for the material costs, with the result that the original forecasting calculation carried out by Contractor was no longer reflective of the actual increased costs experienced on the project. These two events were:

- The extraordinary levels of inflation experienced in most materials prices for the period of this contract, from 2005 to the end of 2009. These high levels of inflation were particularly prominent in the period from 2006 until 2009, a period where the international market had seen a boom in the market of commodities and construction.
- The unexpected delays and disruptions that occurred during the life of the project as a result of unforeseen events, such as a delay in handing over the site and design discrepancies, which were completely outside the control of the Contractor.

Unrealistic and unenvisaged increases were recorded in the prices of all construction materials, including steel, copper, concrete and most other construction-related material components, such as timber and cement, especially until the end of 2009. During that period of exponential boom, the Far East and the Middle East Gulf countries, in particular China, Dubai and the rest of the United Arab Emirates, demonstrated an unquenchable demand for all building materials, and as a result placed a huge burden on supply, with the knock-on effect that prices of particular materials had increased in value in certain instances in excess of 100% during the period from 2005 until 2009.

## 8.3. Contract price

The Contractor considers that the Client needs to aid the return of price equilibrium to the Contract by compensating him for the totally unforeseeable increases in the price of various construction materials.

Because of the exceptional circumstances in respect of inflation experienced in the United Arab Emirates (UAE) during the period in question, the UAE government introduced legislation in the form of a Statute, which provided specific compensation measures to offset the losses incurred by contractors executing projects in the United Arab Emirates. The right of contractors to claim for an increase to reflect the relevant material price escalation was first examined by Statute No. 2072/mb dated 18 March 2008, which called for the formation of a ministerial committee to study the problems facing contractors in the implementation of projects, and to look at how to deal with the phenomenon of exceptional inflation. This ultimately resulted in the Decree of the Councils of Ministers No. 155 dated 9 June 2008, which was taken into law.

The said law took into consideration compensation for exceptional price inflation on construction materials (including reinforced concrete, cement, steel and timber) relating to construction projects that were executed or under execution after 20 January 2005, and which were based on bills of quantities described in and forming part of the applicable construction contracts.

Decree No. 155 outlined the rules and the procedures to deal with increases in cost due to material price inflation relating to the execution of construction projects. A joint committee was formed from the Ministries of Water, Electricity, Education,

Transportation, and Health and Municipality to provide a clear and precise mechanism to compensate contractors for increases in prices based on a contract's classification and scope. The committee decided on the average increase in material cost, based on indexes provided by the necessary authorities for particular periods between the years 2005 and 2009.

Pursuant to the above, the Contractor relies on Clause C.1 Paragraph 2 of its Contract with Client, where its Contract Price is not limited on 'Government-caused cost increases imposed at any time':

**Clause C.1 Paragraph 2.**
Except as otherwise provided herein below, the Contract Price constitutes the entire compensation due to the Contractor for the Work and all of Contractor's obligations hereunder, delivery and completion of Work at the site . . . (including design wherever required under the Scope of Work) but does not include for compensation for any government-caused cost increases imposed at any time, or for any applicable taxes, duties, Customs Duties, fees, overheads, profit, mobilisation and demobilisation not known about at the Base Date of the Contract.

## 8.4. Compensation claim

The amounts included in this Statement of Claim are submitted in accordance with the compensation mechanisms outlined in Decree No. 155, as referred to above.

On that basis, the Contractor is claiming reimbursement in the total amount of $111 109 521 (as summarised in Table 8.1).

The Contractor, under this claim, requests the payment of his direct and indirect costs in relation to: materials costs as they relate directly to the procurement of these materials; shipping costs, which also has increased due to the high cost of petrol; insurance costs; letter of credit fees and bank finance charges; and transportation of materials to site. These items have all been affected by the unprecedented increase in costs, and the Contractor has relied on his Bill of Quantities, which included percentage fees for

**Table 8.1** Summary of the amount claimed

| Section | Amount due: $ |
|---|---|
| Section 01 | 27 633 792 |
| Section 02 | 12 002 656 |
| Section 03 | 25 996 439 |
| Section 04 | 30 984 088 |
| **Subtotal** | **96 616 975** |
| Direct and indirect costs, and financial charges: 15% | 14 492 546 |
| **Total** | **111 109 521** |

direct and indirect fees, and hence added 15% to the increased cost of materials for compensation against these fees (as shown in Table 8.1).

## 8.5. Calculation methodology

The following steps were taken to calculate the material cost per Section for the years 2006, 2007, 2008 and 2009 during which time the procurement of the materials was undertaken

- first, the total amount of each Section in the Bill of Quantities was calculated
- then, the percentage of each applicable material cost per total amount per Section was established by Contractor at the time when the Bill of Quantities was prepared (see Table 8.2)
- this was then converted into a financial value for each material (see Table 8.3).

The dates and rates of execution for each Section as set out in the monthly Progress Reports prepared by the Contractor, which were approved by Client, were then established, as set out in Table 8.4.

The rate of progress of each Section was then applied to the value of the applicable materials to show the expenditure on each material type for each Section in each of the years from 2006 to 2009. This is shown in Table 8.5.

**Table 8.2** Percentage of each applicable material cost per total amount per Section at the time when the Bill of Quantities was prepared

| Section | Contract amount: $ | Percentage of applicable material cost | | | |
|---|---|---|---|---|---|
| | | Reinforced concrete | Cement | Steel | Timber |
| Section 01 | 169 600 000 | 24 | 12 | 21 | 18 |
| Section 02 | 105 400 000 | 20 | 10 | 12 | 10 |
| Section 03 | 90 800 000 | 21 | 18 | 24 | 21 |
| Section 04 | 152 600 000 | 24 | 12 | 21 | 15 |

**Table 8.3** Converted financial value of each applicable material cost per total amount per Section

| Section | Contract amount: $ | Converted financial value: $ | | | |
|---|---|---|---|---|---|
| | | Reinforced concrete | Cement | Steel | Timber |
| Section 01 | 169 600 000 | 40 704 000 | 20 352 000 | 35 616 000 | 30 528 000 |
| Section 02 | 105 400 000 | 21 080 000 | 10 540 000 | 12 648 000 | 10 540 000 |
| Section 03 | 90 800 000 | 19 068 000 | 16 344 000 | 21 792 000 | 19 068 000 |
| Section 04 | 152 600 000 | 36 624 000 | 18 312 000 | 32 046 000 | 22 890 000 |

**Table 8.4** Dates and rates of execution for each Section

| Section | Rate of execution: % | | | |
|---|---|---|---|---|
| | 2006 | 2007 | 2008 | 2009 |
| Section 01 | 50 | 30 | 20 | 0 |
| Section 02 | 30 | 50 | 20 | 0 |
| Section 03 | 20 | 20 | 50 | 10 |
| Section 04 | 0 | 20 | 20 | 60 |

The 'inflationary' percentage increase in the cost of concrete, cement, steel and timber, as set out in Decree No. 155 and as established by the Chamber of Commerce Committee, is given in Table 8.6.

The inflationary price increase percentage (as set out in Table 8.6) was then applied to the appropriate material value in each period and for each Section (as set out in Table 8.5), and this resulted in the inflationary price increase set out in Table 8.7.

**Table 8.5** Expenditure on each material type for each Section

| Section/year | Expenditure (amount in contract): $ | | | |
|---|---|---|---|---|
| | Reinforced concrete | Cement | Steel | Timber |
| **Section 01** | | | | |
| 2006 | 20 352 000 | 10 176 000 | 17 808 000 | 15 264 000 |
| 2007 | 12 211 200 | 6 105 600 | 10 684 800 | 9 158 400 |
| 2008 | 8 140 800 | 4 070 400 | 7 123 200 | 6 105 600 |
| 2009 | 0 | 0 | 0 | 0 |
| **Section 02** | | | | |
| 2006 | 6 324 000 | 5 270 000 | 6 324 000 | 5 270 000 |
| 2007 | 10 540 000 | 3 162 000 | 3 794 400 | 3 162 000 |
| 2008 | 4 216 000 | 2 108 000 | 2 529 600 | 2 108 000 |
| 2009 | 0 | 0 | 0 | 0 |
| **Section 03** | | | | |
| 2006 | 3 813 600 | 3 268 800 | 4 358 400 | 3 813 600 |
| 2007 | 3 813 600 | 3 268 800 | 4 358 400 | 3 813 600 |
| 2008 | 9 534 000 | 8 172 000 | 10 896 000 | 9 534 000 |
| 2009 | 1 906 800 | 1 634 400 | 2 179 200 | 1 906 800 |
| **Section 04** | | | | |
| 2006 | 0 | 0 | 0 | 0 |
| 2007 | 7 324 800 | 3 662 400 | 6 409 200 | 4 578 000 |
| 2008 | 7 324 800 | 3 662 400 | 6 409 200 | 4 578 000 |
| 2009 | 21 974 400 | 10 987 200 | 19 227 600 | 13 734 000 |

**Table 8.6** Inflationary percentage increase in the cost of concrete, cement, steel and timber

| Section | Inflationary increase in cost: % | | | |
|---|---|---|---|---|
| | 2006 | 2007 | 2008 | 2009 |
| Reinforced concrete | 4.86 | 18.18 | 24.00 | 28.25 |
| Cement | 18.05 | 30.58 | 32.79 | 24.97 |
| Steel | 18.64 | 29.45 | 107.43 | 20.80 |
| Timber | 6.75 | 12.50 | 18.65 | 17.65 |

**Table 8.7** Inflationary price increase

| Section | Price increase due to inflation: $ | | | | |
|---|---|---|---|---|---|
| | Reinforced concrete | Cement | Steel | Timber | Total |
| **Section 01** | | | | | |
| 2006 | 989 107 | 1 836 768 | 3 319 411 | 1 030 320 | 7 175 606 |
| 2007 | 2 219 996 | 1 867 092 | 3 146 674 | 1 144 800 | 8 378 562 |
| 2008 | 1 953 792 | 1 334 684 | 7 652 454 | 1 138 694 | 12 079 624 |
| 2009 | 0 | 0 | 0 | 0 | 0 |
| **Section 01 total** | | | | | **27 633 792** |
| **Section 02** | | | | | |
| 2006 | 307 346 | 951 235 | 1 178 794 | 355 725 | 2 793 100 |
| 2007 | 1 916 172 | 966 940 | 1 117 450 | 395 250 | 4 395 812 |
| 2008 | 1 011 840 | 691 213 | 2 717 549 | 393 142 | 4 813 744 |
| 2009 | 0 | 0 | 0 | 0 | 0 |
| **Section 02 total** | | | | | **12 002 656** |
| **Section 03** | | | | | |
| 2006 | 185 341 | 590 018 | 812 406 | 257 418 | 1 845 183 |
| 2007 | 693 312 | 999 599 | 1 283 549 | 476 700 | 3 453 160 |
| 2008 | 2 288 160 | 2 679 599 | 11 705 573 | 2 288 160 | 18 961 492 |
| 2009 | 538 671 | 408 109 | 453 274 | 336 550 | 1 736 604 |
| **Section 03 total** | | | | | **25 996 439** |
| **Section 04** | | | | | |
| 2006 | 0 | 0 | 0 | 0 | 0 |
| 2007 | 1 331 649 | 1 119 962 | 1 887 509 | 572 250 | 4 911 370 |
| 2008 | 1 757 952 | 1 200 901 | 6 885 404 | 853 797 | 10 698 054 |
| 2009 | 6 207 768 | 2 743 504 | 3 999 341 | 2 424 051 | 15 374 664 |
| **Section 04 total** | | | | | **30 984 088** |

The Section totals, i.e.

| | |
|---|---|
| Section 01 | $27 633 972 |
| Section 02 | $12 002 656 |
| Section 03 | $25 993 439 |
| Section 04 | $30 984 088 |

are then transferred to the summary table (Table 8.1).

## 8.6. Conclusion

With respect to the increased costs of materials, the Contractor accepts that the Contract generally transferred the risk of price increases to the Contractor during the Contract Period. To price this risk, the Contractor executed extensive research on the price increase trends during the 5-year period leading up to the submission of the tender, and this price risk was included in the tender.

However, the actual price increases that occurred were totally unforeseeable by the Contractor. Because of the extraordinary inflation rates applicable to the materials in question, and with a view to alleviating financial hardship caused by these unforeseeable and catastrophic price increases, the government of the United Arab Emirates intervened and introduced legislation in the form of Decree No. 155, which provided specific compensation measures to offset the losses incurred by Contractors executing projects in the Emirates, especially during the period 2005 to 2009 when the international markets had experienced exceptional increases in the price of certain construction materials. Contractors' position for a claim for an increase to reflect the said material price escalation has been examined by Decree No. 2072/mb dated 18 March 2008, which called for the formation of a ministerial committee to study the problems facing contractors in the implementation of projects due to materials cost increases and to look at how to deal with this phenomenon. This ultimately resulted in the Decree of the Councils of Ministers No. 155 dated 9 June 2008, which has taken into law the following: 'The law takes into consideration the compensation for works that have been executed or under execution after 20/1/2005 based on quantities described in Contract Documents and Bills of Quantitative described by the Consultant and approved by the Designated Party.'

The works that this Decree includes are 'Reinforced Concrete, Cement, Steel and Timber'.

Based on Decree No. 155 stating the acceptance of the rules and the procedures to deal with delays and increases in cost due to material escalation, the Contractor has set out above its entitlement to the applicable increased costs.

The Contractor relies on his contractual and legal right for complete compensation of all losses he incurred due to the increased costs of materials at Project B. These costs have placed the Contractor under extreme duress, and could have made the Contract annulled under similar conditions if the works were to be carried out by another contractor with limited financial capabilities. The Contractor is not attempting to make any profit or compensation by submitting this claim, but merely seeks to recuperate losses (which it is entitled to do in line with statute) that it incurred due to these extreme and unprecedented increases in cost.

The Contract between the parties permitted the recovery of these said costs, and the Contractor seeks as compensation in respect of this matter the sum of $111 109 521 (as set out under Table 8.1).

**Delay and Disruption Claims in Construction**
ISBN: 978-0-7277-5705-0


doi: 10.1680/ddcc.57050.107

# Index